Game Design Theory

Game Design Theory

A New Philosophy for Understanding Games

Keith Burgun

CRC Press
Taylor & Francis Group
Boca Raton London New York

CRC Press is an imprint of the
Taylor & Francis Group, an **informa** business

AN A K PETERS BOOK

CRC Press
Taylor & Francis Group
6000 Broken Sound Parkway NW, Suite 300
Boca Raton, FL 33487-2742

© 2013 by Taylor & Francis Group, LLC
CRC Press is an imprint of Taylor & Francis Group, an Informa business

No claim to original U.S. Government works

International Standard Book Number: 978-1-4665-5420-7 (Paperback)

Library of Congress Cataloging-in-Publication Data

Burgun, Keith.
    Game design theory : a new philosophy for understanding games / Keith Burgun.
      p. cm.
    Includes index.
    Summary: "This work looks at how digital games fit into the long history of games and offers solutions to some of video games toughest design challenges. It covers the art and craft of developing a set of rules to create a contest between players or other agents, targeted specifically at digital game designers. The author outlines a somewhat radical philosophy based on thousands of years of game design, illustrating how we must adhere to these ancient principles if we are to improve digital games in the future"-- Provided by publisher.
    ISBN 978-1-4665-5420-7 (pbk.)
    1. Computer games--Design. 2. Computer games--Programming. 3. Video games--Design. I. Title.

QA76.76.C672B86 2012
794.8'1536--dc23
                                          2012019337

Visit the Taylor & Francis Web site at
http://www.taylorandfrancis.com

and the CRC Press Web site at
http://www.crcpress.com

Table of Contents

6. Predictions 151

Index 163

Foreword

S ome recent research suggests that by the age of 20, the majority of Americans have spent as much time playing video games as they have spent time at school—and I guess other countries are catching up fast. I find this remarkable! Others may find it frightening....

Certainly, these findings represent a great challenge (and a great responsibility) to those of us who are game designers. If we can reach and influence so many people with our games, what are we doing with this influence?

In recent decades, games have become increasingly popular and have grown to be a significant market force. The emergence of powerful video games has boosted the popularity and attractiveness—some call it addictiveness—of games as a meaningful pastime. Today the games industry is larger than either the movie industry or the music industry, and games now compete with books for the top revenues in the entertainment business. As a member of the games industry, I find these developments remarkable too (although members of the other industries may find them unnerving).

This revolution goes far beyond the traditional scope of playing games. Our smartphones offer us a half-million games at our fingertips, many of them for free. Games have taken social communities, such as

Facebook, by storm, involving millions of players in a single gaming experience. The online role-playing game *World of Warcraft* alone attracts millions of players, who collectively have spent more than six million years on the game—and counting. This is comparable to mankind's total global effort in putting a man on the moon!

Of course, the amazing success of games has not gone unnoticed by the rest of the world. Today we can see how the attractiveness of gaming elements has resulted in them being applied to many areas of our lives: enticing incentive programs, motivating fitness programs, and ever-present leaderboards are all popular manifestations of this "gamification" process. Games have become truly global!

However, despite the rise of games and gamers, the creative game design process remains largely unstructured. Game designers are often self-taught, or serve apprenticeships under more experienced designers. They each develop their own methods of design, their own vocabularies, and their own toolboxes of tricks to identify and fix problems. Unlike literature and music, which stand on solid theoretical foundations, game design theory is much less developed. Game designers are artists, and each has his or her own philosophy of how to squeeze the most fun and enjoyment out of a game box.

It is possible that thought-provoking books such as this one may be just the spark required to kick start an industry revolution in game design.

—Reiner Knizia
London, England
March 2012

Introduction

The Death of *Tetris*

You don't need to be an expert on the topic of games to have a sense of the level of elegance, brilliance, and importance of *Tetris*. An abstract, score-based game based on fitting various four-block shapes (known as *tetronimoes*—or *tetriminos*, in the parlance of *Tetris*) into each other to create lines (filled horizontal lines that go across the *well*, or playing field) took the world by storm in the mid-1980s, exploding even further with the release of the Nintendo Game Boy version in 1989.

What makes *Tetris* so brilliant? With so few gameplay elements, it would seem as though the game would be simple and mastery would be easy, but that's far from the case. *Tetris* has achieved the game design feat of "easy to learn, difficult to master" more than most video games—it is incredibly intuitive to learn, and yet I've been playing it for over 20 years and I am still learning things all the time.

The depth of *Tetris* is found in several aspects of its gameplay, but two specific areas stand out. The first is learning about relationships between pieces and pile shapes: for instance, you often can use an L-tetrimino in a somewhat nonintuitive way to help you build towards clearing four lines at once—a *tetris* (Figure 1).

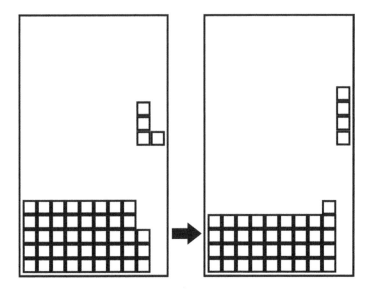

Figure 1. An example of a nonintuitive, yet strategic move in *Tetris*. New players may not realize that this is a solid way to set themselves up for a tetris.

The second, even more significant area of mastery is risk management. You see, *Tetris* generates random tetriminos each time, and so there are often times when you have to make a "push your luck" sort of decision in order to get a better score. For instance, take a look at the scenario in Figure 2.

In the situation illustrated in Figure 2, you could make the safe play and flip the L-piece twice so that it fits in and gives you a nice, safe triple that provides a little wiggle room. The downside, however, is that you lose an opportunity for a tetris, which is worth far more points. The points you lose will be even greater if you're at a higher level (which may well be the case, given that the pile is so high). So, you can *choose* to push your luck by making the play from Figure 1 and waiting for the line piece you need. The thing is, due to the random generator you don't know exactly when that line piece will be coming—it may be two pieces away, or it might be thirty pieces away, and you have no way of knowing! This randomness means you constantly have to adapt to the system, making the outcome of decisions more uncertain.

Perhaps some readers will say, well, at that height I would certainly go for the triple and go into clean-up mode. That's reasonable. But what if the pile was two tiles lower than it is in Figure 2? What if it was three or four tiles lower? There is no firm line at which a player *must* begin to play it safe, and sometimes taking a big risk has a big payoff.

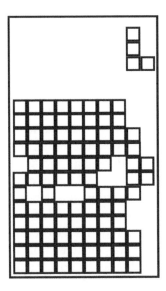

Figure 2. A higher stakes situation in *Tetris*.

Tetris was packaged with the Nintendo Game Boy, and for many people the game went into the system when they first got the Game Boy and it remained there. The game's deep, elegant mechanisms combined with its random piece-generator meant that it always had something new to teach—it always was putting players in positions that they hadn't totally learned to deal with yet.

At least, that used to be the case.

Starting around the turn of the millennium *Tetris* started to change. Newer versions added all kinds of features that seemed to do everything in their power to take that ambiguous-decision quality away from the game. Instead of the normal single *next box* (a very helpful user-interface (UI) space that showed which piece was coming up next), we started getting three next boxes—then four, or five. Now many versions have six, meaning that there is absolutely no uncertainty about how the next six moves will play out.

If that wasn't enough, a feature called the *hold box* was added. The hold box allows players to save one piece for later: at any time, players can swap out a current piece for the piece in the hold box. This change almost completely destroys the dilemma inherent in dealing with combinations of pieces and piles that players don't know how to manipulate.

The game takes further abuse from another new feature called *easy spin*. Although this doesn't directly affect the decision-making aspect of the game, it does remove the element of tension that goes with each piece having a "timer." Easy spin allows players to spin a piece at the bottom of the well *indefinitely*, giving them unlimited time to decide where to actually put that piece.

But possibly one of the most offensive and least talked about changes is how the random generator works. Early versions of the *Tetris* generator either worked completely randomly, or had a very slight cap on repetition so that you wouldn't get the same piece ten times in a row. But now there is something called the *7-Bag*, which works by putting all seven tetrimino possibilities into a bag and drawing them one at a time. This system guarantees that you will get one of each piece every seven pieces, and makes piece generation completely regular and dependable. It's funny how the modifying of such a behind-the-scenes, small mathematical algorithm can completely change the nature of a game, but that's what happened. This feature was the final mortal blow to any uncertainty in decision making, and it shows just how fragile a game really is.

These new features have added up to a new reality: that decision making in modern *Tetris* is actually pretty trivial. Instead modern *Tetris* has become more of an execution and reaction contest—almost akin to a rhythm challenge like *Dance Dance Revolution*. Today's serious *Tetris* players play versions of the game that fire pieces at incredible speeds (five or more per second). Knowing where to put the pieces is not very important: it's just a matter of *doing it in time*. For those who play the newer games at normal speeds, the game is ridiculously easy and gets boring well before they're ever threatened. Modern *Tetris* isn't even close to being the same game that we fell in love with in the 1980s and 1990s. The original *Tetris* was one of the most important examples of digital game design excellence, and yet today it's very difficult to access or find a version of the game without the new features. How could we let this happen?

The reason is that we never understood collectively what was so great about *Tetris* in the first place. We never "got" the game, oftentimes calling it a puzzle, ostensibly because the pieces fit together somewhat like those in a jigsaw puzzle. We didn't even really know what we meant by *puzzle*, and we didn't know what we meant by *game*—the two terms were often interchangeable. We enjoyed the software but we didn't know *why* we enjoyed it, and now we've made what was great about it inaccessible to a whole generation: a generation that will grow up thinking that *Tetris* is boring. And they're right: the version they have access to is forgettable and lacks those hooks that kept players tied in for so many years. The game that *Tetris* was inadvertently has been lost, and that's why I'm writing this book.

Our Story

Games have always been important to people, but for nearly the entire history of human civilization making games has never been an established craft in the way that music, writing, and the visual arts have been. People have always created games, of course, but until recently there has never been a specific class called *game designer*. We game designers haven't had our Bachs or our DaVincis, people who established guidelines and principles for how our craft really works in a scientific and reproducible way. A sad fact about the world is that if you can't make a living doing something, very few people will pursue it seriously as a craft—so while each culture has followed its own evolution of creating sports, contests, and tabletop games, the evolution has been slow and the understanding superficial.

That changed dramatically in the 20th century. We suddenly find ourselves in an era in which being a game designer is actually a viable way to make a living, probably for the first time in human history. Why has this become the case only recently? One reason is that learning and exploring games takes a lot of time, and until recently people didn't have enough free time to learn a large number of them, limiting the demand for new games. Further, games could afford to be less complicated when free time was more rare.

So here we are—the very first generation of human beings to have been asked to satisfy the sort of demand we're seeing now. How are we doing? Actually, although it's completely understandable given the circumstances, we're in a very unstable, unhealthy, and unsustainable position with respect to how we view and create games. In short, we don't have any kind of established understanding about what games are, how they work, or what they ought to be. We're stuck in a place where all we can say is that some people like some games, and some people like other games. It's impossible for us to engage in any kind of productive discussion or critique of games, and we really can't progress until this problem is solved.

What is the solution? Essentially we're in a dark room, and right now everyone is afraid to reach out and try to touch something. The solution lies in game designers boldly saying *something* about games, in presenting their theories. There are a growing number of designers out there right now who are proposing hypotheses, which is a sign that we should have some optimism about the future of games.

My Story

Like many people, I grew up playing video games. Like slightly fewer people, I continued playing video games as an adult. I became part of

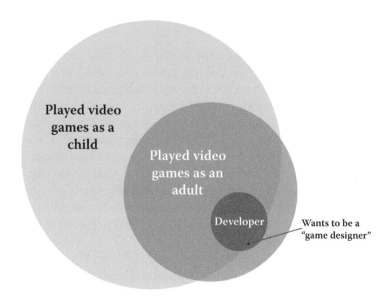

Figure 3. Distillation of a game designer.

an even smaller group when I decided that I wanted to create my own games. Finally, I entered an even tinier circle: I decided that I wanted to specialize in game design (Figure 3).

This last step is not that common among those who live in the digital world of games, and the reasons are clear. Many video gamers get into game development through computer-related disciplines—most commonly coding, since computer programming is the most significant practical aspect of bringing any kind of computer application into existence. Indeed, if you are a person who wants to make games, learning to program is the fastest way to start making that happen.

And that is what I did (sort of). In 1994, when I got my first computer, I immediately started tinkering with QBASIC, a variant of the BASIC language that came bundled with most versions of DOS. I used this language to create lots of little shooter games, fighting games, and other small experimental games. They were all very simple—some even simpler than they really should have been—and many were left unfinished. While I'd like to blame these things on the limitations of QBASIC, the truth is I just never developed a real love for programming: for me, programming was always just a means to an end. Nearly 20 years later, QBASIC is probably still my strongest language for this reason.

At a certain point I recognized something that some other people didn't seem to—creating a game on the computer had two very different parts to it, and only one of them really interested me. Programming is implementation, but deciding *what* to implement was always what interested me. It seemed that a lot of people thought that this "what to implement" thing was trivial: instead they just copied some other game, tweaked one or two rules, and created new content. This formula never satisfied me, as I always felt that game design could be (and should be) something brilliant and fascinating in its own right. I soon found myself spending less and less time worrying about programming, and more and more time writing down rules for my game ideas with a pencil and paper. In hindsight, I don't think I worried much about actually creating these games. What was important was that *I was designing games*.

Problem Statement

The death of *Tetris* is sadly only the tip of a much greater iceberg. History will not look back kindly on the popular digital games of today, which can be seen by looking back even five years. Who is still playing the hit games of 2007—such as *Bioshock, Call of Duty 4,* or *God of War 2*—in 2012? Not a whole lot of people. That number will continue to dwindle quickly over the next few years, and it would surprise me if more than a handful of people even know about those titles 25 years from now. Put simply, game designers generally are focused on creating games that will sell *today*, as opposed to games that will continue to be interesting tomorrow.

That's not the worst of it, though. There are many terribly destructive trends in design that are causing tremendous damage to our designs and leaving players empty-handed. Even our best attempts at creating good games are plagued with features that ruin them, many of which are *expected* in new games. In short, the video games we play and love—even many that we know as the classics—have massive problems that they don't need to (and before you assume that the independent game-development world is immune to these problems, let me tell you that sadly, they are not). I'll describe these problems in detail in future chapters, but if you're reading this with skepticism, ask yourself these questions:

- How interesting are the dungeon puzzles in the *Zelda* games?
- What effect does *quicksave* have on the game-playing experience?
- How good are the stories and writing in video games, really?
- If there's no element of randomness in a single-player game, what does that do to its replay value?

These are far from the only issues, and are simply a few examples that provide a broad idea of the problem. I'll be going into much more detail on all of these subjects later on in the book.

On Game Design

Game design is the development of the most fundamental aspect of a game: rules. It's obvious to most game developers that game design is not programming. It's also obvious that game design is not content creation (things like three-dimensional (3D) modeling, pixel art, music composition, sound effects, etc.). Writing, storytelling, and even character design are also not game design.

Put simply, game design is deciding what the game's *mechanisms* will be. There may be times when the game designer has to have a certain amount of influence over the visual art in a game (if it even has visual art), but that does not mean that visual art is an inherent part of game design. Any discipline requires us to step outside of the field sometimes to get something done. The fact that an architect sometimes has to deal with legal papers doesn't mean that law is a part of the discipline of architecture.

If you're interested in learning about game design, what can you do? Well, there are a number of books out there that you can buy, but nearly all of the game-design books I've seen are at the introductory level. It's very hard to finds books that are more than loose, general, safe introductions to the art of game design. To understand game design, you need to read (and maybe even write) game design books with a *philosophy* behind them, but unfortunately, most of the books, blogs, and articles that are available steer clear of actually saying something bold about games.

My book is not that kind of book. I *do* have a point of view, and I think one of the things we need at this time are books that carefully illustrate new hypotheses on designing games, not ones that simply state that all thoughts on the matter are equally valid. No serious physicist reads *Physics for Dummies*: they read works that inhabit the cutting edge of understanding, that strive to further our comprehension of the subject. I want to read a game design book that has something to say. The fact is, game designers deal with very deep, very difficult concepts about the workings of human beings that ultimately no one has the answer to. Game design is an exploration, and we designers should have the courage to explore.

We need game design movements driven by a design philosophy. I'm not talking about genres or other, more superficial classifications. A quick look at art history yields examples of what I mean: realism, expressionism, dadaism, and cubism were all catalyzed by artists who had a real point of view about what art *should be*. It's about time that we in game design started to have the same kind of serious conversation. I reject the idea that everyone's opinion is equally correct—I think that there are real answers about what games are, and how they work, waiting to be discovered. We just have to try.

Game Design Theory Today

Some may respond that this conversation has already been taking place. Well, yes and no. Over the past decade or so, a number of working game designers have written books about game design. Unfortunately, most of these are not about game design at all, but instead give advice on the practical aspects of game *development*. Some are essentially programming books, some are focused on making it in the industry, and some address other tangentially related topics. The number of game design books that are actually about game design is much smaller. Such books do exist, but I have yet to come across one that puts forth a bold vision: a philosophy of what games are and what they should be.

Challenges for Game Designers, written by Brenda Brathwaite (of Sir-Tech fame) and Ian Schreiber, is a popular book on game design that is also a great example of the problem I see with these books. Much of the book is pretty basic introductory textbook–type stuff, and although it includes hands-on exercises (which are useful), the book slams on the brakes anytime it comes close to talking about design philosophy. For instance, there's a section titled "Narratology and Ludology." According to the authors, *ludology* is "the study of games as rules (or mechanics)" and *narratology* is "the study of games as a storytelling medium." This short section ends with the following statement:

> These two divergent schools of thought are, for the most part, exactly that—thought. In the life of a workaday game designer, the topics are rarely discussed in black-and-white definitions as they are above. Rather, the designer usually focuses on what's not up to snuff in the game, whether it's something whacked with the balance or an untested story path he has yet to implement.

The authors end the section by saying that "the two schools are complimentary" without any explanation of how they are complimentary, and completely overlook all of the times that the schools are anything *but* complimentary. In my opinion this is an attempt to be as safe and conciliatory as possible, which ends up being a complete waste of text. Why bother writing down this standard-issue, status quo half-opinion? How is this chunk of text useful for anyone? What does "up to snuff" even *mean*?

Another well-known book on game design is Raph Koster's illustrated book, *A Theory of Fun for Game Design*. While this book does make some solid points, he stops short of having a holistic, complete vision for what games are. For instance, he has a chart on one page showing numerous different human activities—all kinds of things, from "community" to "performance" to "criticism" to "teaching." He then goes on to say this:

> The classic definition of *game* covers only some of the boxes in the grid. Arguably, all of the boxes in the grid are fun to someone. We need to start thinking of games a little more broadly. Otherwise, we will be missing out on large chunks of their potential as a medium.

He is essentially saying that just because someone thinks something is fun (and "someone" can think *anything* is fun), we need to expand the definition of the word *game* (already extremely loose, if you ask me) to include whatever that activity is. This kind of talk moves us further from a solid understanding of what games are, not closer to it.

Other examples include Jesse Schell's *The Art of Game Design: A Book of Lenses*, which has exactly 100 lenses, or questions, to ask yourself about your game design. First, what are the chances that there are exactly 100 good questions that need to be asked? (I propose that there should be a *rule of suspiciously round numbers* that tells us to doubt such lists.) Overall the book may end up having some use for game designers, but it's definitely using a "spray and pray" approach, since it's likely that only two or three of these questions will actually be useful. Again, it does *not* provide a holistic view of the nature of games and will only improve your understanding of them circumstantially.

For those who might defend these books by saying that they're only giving readers wiggle room, or that they're allowing readers to come to their own conclusions about what games are: *readers do not explicitly need to be given permission to do this.* Thinking persons will come to their own conclusions, regardless of whether they read something wishy-washy, or something pointed. One can make a strong point and still allow disagreements and other ideas to exist.

The essential problem with game design theory now is that too many people are resistant to any solution that may be a little bit *destructive*. "If a solution means I have to throw the gameplay of *Final Fantasy VII* into question, then forget it!" might be one reaction. Language and culture may also be impediments to change: the meaning of the word *game* is very broad and very loaded culturally. We may need terms that are more specific than those that are currently available if we are ever to get a clearer understanding of the different types of interactive systems.

What This Book Is

This book is a walk-through of my philosophy on game design. It offers a radical yet reasoned way of thinking about games, and a holistic solution to understanding the difference between games and other types of interactive systems. I argue that the method offered in this book is *the* path that game design must take to improve.

I propose definitions, concepts, and methods that together form a philosophy of game design. This book aims to add this philosophy to the ongoing discussion in a bold and clear way. Even if you completely disagree with what you read here, you will certainly come away with a stronger understanding of the field and a more distinct philosophy of your own, which will make you a better game designer. After laying out the fundamental concepts of my philosophy, we'll use it as a lens to analyze the history of games and modern trends.

This is a book for people who, like me, wish to find the best way forward for games.

What This Book Is Not

Some of the game-design resources I've looked at go on at length about the cultural *meaning* of games in our society. They discuss the games industry, the state of gaming journalism, the role of race and gender in games, "gamification," and other topics loosely related to game design. General statements about the experience of players and the nature of play are also common. Almost all of them seem to downplay, minimize, or outright ignore the purely mechanical aspects of games, which I think is a serious problem that has affected games in a profoundly negative way.

As I've already made clear, this is not a hands-on, how-to book about game design or game development. It's not about how to sell more copies of your game, and it's not about how to work better with a team. Those things are absolutely useful to any commercial game designer, but they aren't so useful for people who just want to make a small game with pen and paper to play with their friends—let alone people who just want to design, and don't even want to play the games they create.

This is a book that will be useful to *all* game designers, because it is a book about game design at an abstract and fundamental level. It is specifically directed at video-game designers and players. As I said before, I think video games and the culture surrounding them are in a very unhealthy place right now, but at the same time video games have enjoyed incredible success over the last ten or fifteen years. Consequently, designers have even more responsibility for knowing what they're doing! Beyond that, though, we as game designers now have possibilities available to us that were never available before. We can play chess with people on the other side of the world. We can play highly complex games in real time. We have online leaderboards and rankings for score-based games and tournaments. And we can provide balance patches with the flick of a switch and offer extra content over the Internet. The possibilities offered by the digital realm of gaming are magnificent, but we have not been taking advantage of them. This is why I am so focused on video games: I see massive potential that is currently being squandered.

Rest assured: designers of board games, athletic games, or any other kind of games also will get a lot out of this book, as its lessons apply to all kinds of games. My explanations reference all types of games, and not just digital ones.

Finally, let me be perfectly clear—my aim is to help as many people as possible create games that are as much fun as possible. This book is a manifesto of how I think that can be done.

1

The Concept of Game

In my view, the very breadth of the term *video game* has had a profoundly negative impact on the way that we in the video-game world think about games and game design. The term has become a catchall for any kind of digital interactive entertainment, regardless of the nature of the system. Furthermore, for most people who play video games, the massive catchall term *video game* has become synonymous with *game*. And yet puzzles, sandboxes, toys, simulators, interactive fiction, contests, and many other types of entertainment also are referred to as games. The problem with this is that there actually is a type of system that's getting lost in the mix—a specific thing that is *not the same as* a puzzle, a toy, or a basic interactive system such as a *simulator*. The effect of using the term *game* for all of these things is that we're left with no way to point to this other, unique type of system.

This question of words is really very significant. How can we ever develop guidelines to make better games if we're calling both *The Path* and chess *games*? What do these two things even have in common, besides the very vague idea that they're both interactive?

In this book I propose a prescriptive definition for the word *game* that allows it to fit in nicely with other types of interactive systems, such as puzzles and contests. Additionally, this definition makes it clear that games have a unique identity that is different from other types of interactive systems. I should mention that this is not some strange, new, arbitrary definition: many game designers already agree with my proposed usage in a general way. Furthermore, I think that it's actually very consistent when you look at the way *game* is applied to games that aren't digital. Very few people refer to jigsaw puzzles as games; indeed, jigsaw puzzles get their own area in the store.

Definitions

Merriam-Webster's first definition on its list for the word *game* appears below.

> Game: activity engaged in for diversion or amusement.

This definition is perfectly fine for everyday use. The dictionary is doing its best to cover all bases, and as a result the first definition is exceptionally broad. For those of us who are serious about the subject of games, however, this definition is woefully inadequate. It implies that eating a hamburger or watching television could be considered games, and I think even most laypeople would consider it too broad. The third definition on Merriam-Webster's list is much closer to one that is useful to us.

> Game: a physical or mental competition conducted according to rules with the participants in direct opposition to each other.

This definition is quite close to what I think a game actually is. It includes the aspect of *competition*—there are different *agents* trying to achieve a goal that cannot be shared. It mentions *rules*—guidelines by which the game must be played. These are both important features of this definition (ones that I'll address in more detail later), but there is something very important missing from the definition above.

First, though—do you think a weight-lifting contest is a game? How about a hot-dog eating contest? An arm-wrestling match? Some of you will answer no, and those of you who do not will at least hesitate before defining these things as games—if a friend asks whether you want to play a game and then reveals that he wants to have an arm-wrestling match, part of you will be surprised. The idea of *playing a game* feels like it should involve something more than merely measuring the strength of your arm against that of an opponent.

So what is that missing element? What makes contests different from something like chess or *Street Fighter,* both of which we consider games?

Let's look at my definition of game to find out.

> Game: a system of rules in which agents compete by making ambiguous decisions.

The act of ambiguous decision making is what's really missing from that dictionary definition, and it is what makes a game a special *type* of contest. Since there's little debate about what the word *contest* means, we can use that to shorten my definition.

> Game: a contest of ambiguous decision making.

Throughout this book, I will be using *game* to reference the concept above. I'll be using the term *video game* to refer to a whole package—such as the complete software of *The Legend of Zelda*—which includes the system of rules any game has, plus the art, music, theme, and other features that make up video games as we know them.

I'd like to be clear that I'm not saying people are wrong for calling an arm-wrestling match, *Whac-A-Mole*, or *Dance Dance Revolution* games. There are many definitions for the word, and people are certainly correct if they're using the dictionary's first definition. My definition is one that I'm proposing for game designers and game critics, to help them understand games in a clear and more consistent way. Feel free to come up with your own word for this kind of system, though—the important thing is that we have a solid understanding of the *concept* of games as I have described it.

Mapping Interactive Systems

The chart in Figure 4 illustrates the starting point for our philosophy. You'll need to understand the chart and the definitions associated with it for the rest of the book to make sense: much of what follows uses this chart as a lens through which to examine games, in order to develop a better understanding of how they work.

Interactive Systems

- Obvious examples: *SimCity, Microsoft Flight Simulator*
- Less obvious examples: *Dwarf Fortress, Minecraft*

All of the categories mapped in Figure 4 are interactive systems, which can be defined as *possibility spaces defined by explicit rules*. Everything in life is really an interactive system, so this is an extremely broad description. Something that is only an interactive system and not a puzzle,

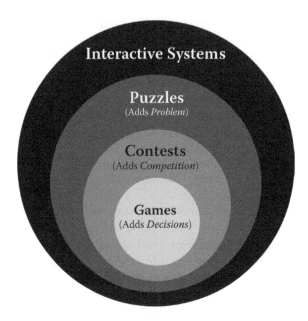

Figure 4. A map of the types of interactive systems: everything on the chart is an interactive system, but not everything is a game. For example, a contest is a type of puzzle, which is a subset of interactive systems, but not included in the game category. A game is an interactive system with all three of the features shown: problems, competition, and decisions.

contest, or game could also be referred to as a *toy*, but that shouldn't imply that basic interactive systems are only for children. Simulators are also basic interactive systems.

Video games such as *Dwarf Fortress* and *Minecraft* are sometimes erroneously called *games*. Some claim that the implicit goal in these applications is survival—but if that's the case, at what point have you "survived" and won? If the way that you "win" is by surviving, then this is a system that cannot be won, because survival is never a condition that is fulfilled. There is no point at which victory is achieved—therefore, these titles are not *contests*, and being a contest is an important part of being a game. This point may be confusing, since players often add their own win conditions to these applications. When they do this, they make a game out of *Minecraft* in the same way that one makes a game out of a flight simulator or Legos. In these cases, a player has actually taken on the role of game designer! *Minecraft* itself did not ship with any rules for win conditions, and so the conditions that players add are not an inherent feature.

Puzzles

- ◆ Obvious examples: An individual level in *Portal*, a problem in *Professor Layton*, a dungeon in *Ocarina of Time*
- ◆ Less obvious examples: An individual level in *Super Mario Brothers* (sort of)

The first circle inside of the one for all interactive systems represents *puzzles*. Essentially, a puzzle adds a *problem*, which of course has a solution. Another way of putting it is that a puzzle adds a *goal*. Keep in mind that the word *puzzle* conveys a certain sense of being difficult or brain bending, but something doesn't need to be difficult to be a puzzle. A dead easy puzzle is still a puzzle (perhaps just not one that many would consider good).

The reason that a level of *Super Mario Brothers* is only roughly a puzzle is that it has no random elements, and therefore has only one optimal solution. Programmed speedruns[1] of the game clearly illustrate this point. Nearly all single-player video games that have no random elements are puzzles only in the same way that *Super Mario Brothers* is a puzzle.

Also less obvious to many people is that there is a category of interactive systems that sometimes are called puzzles, which really aren't puzzles at all, but games. Video games such as *Tetris*, *Dr. Mario*, and *Bejeweled* are often called *puzzle games*, probably because they involve putting pieces together. Using our lens, these are actually games because of their random elements and score-based systems. These kinds of games are not about finding solutions.

A puzzle is essentially a problem that must be solved: when a puzzle is being designed, it is being designed with a solution in mind. It is not a competition, and it is certainly *not* a game. After a puzzle is solved it usually stops having value to a person, much like a riddle whose answer is already known.

Contests

- ◆ Obvious examples: A weight-lifting contest, a hot-dog eating contest, an arm-wrestling match, *Whac-A-Mole*
- ◆ Less obvious examples: *Guitar Hero* (or almost any rhythm game), most pure racing games, many real-time games roughly qualify

Contests add the element of *competition* to puzzles (which added the element of *solution* to interactive systems). A good way to think about

---

[1] Speedruns are attempts to optimize the play through nonrandom digital games. They are often created using software that allows a user to program actions over time in order to remove any element of human error so that the precise solution can be found.

competition is to think of a puzzle with a solution that a number of agents are attempting to find, or reach. This solution cannot be shared among the agents, however—once one of them finds it, the others lose. The element of competition is the most important feature separating contests from puzzles.

Competitions are won or lost, whereas puzzles are solved or not solved. That's a big difference, since winning means that other agents lose. In a way, competitions sort participants by superiority. For instance, if there are two people working to solve a 500-piece jigsaw puzzle together, the solution is shared by both. There's no way for these people to lose against the puzzle, or against each other—and if they choose to stop solving the puzzle or get stuck, they still have not lost to the puzzle. In contrast, if you give each of the two players the same 500-piece jigsaw puzzle and tell them that the first one to complete the puzzle wins, then it becomes a contest.

Contests also do not have to feature parallel achievements among its agents; there are some competitions in which one party has to achieve one goal before another party achieves a completely different goal. In most cases, though, all conditions cannot be met simultaneously, and so these kinds of contests are still competitions.

Moreover (and this is not specific to contests), agents do not have to be human: one agent can be the game system itself, as in the cooperative board game *Pandemic* (which, by the way, is indeed a game). In this game, between two and four players work together to save the world from four deadly viruses that are threatening to destroy humanity. There are different victory conditions for the viruses and the humans. The first to meet their victory conditions win, and the winner is always either all of the human players or the viruses.

Both agents can actually even be the same human being, as is the case in a game with a high-score system. When you play *Tetris*, *Galaga*, or *Dungeon Crawl: Stone Soup* with the objective of trying to beat your previous high score, you're actually competing against yourself. This is also the case for some racing games (*Super Mario Kart* is a good example) that allow you to compete against your ghosts, which are precise recordings of your performance. And many of us use contest-type systems to improve our workouts. (Can you beat last week's 30 push-ups this week? If you can, then you win; if you can't, then you lose.)

Contests are also usually simpler than games or puzzles, and quite often have a time, strength, or dexterity element. Knowing what the solution is usually isn't part of a contest, although it can be, as it is in an Easter egg hunt.

Games

- ♦ Obvious examples: *Team Fortress 2*, *NetHack*, Football, Chess
- ♦ Less obvious examples: *Tetris*, *Dr. Mario*, almost any Match 3 game

Finally we've reached games. Games are interactive systems that have the *problem* quality of puzzles, the *competition* quality of contests, and another new attribute that makes them very special: *ambiguous decisions*. The kinds of decisions we make in games are unlike any we've experienced so far with the other interactive systems mapped in Figure 4.

Playing games is an art. The decisions you make in a game are special because even if you win, you cannot say for sure that the decisions you made were the correct ones. Other decisions may have blocked your opponent more effectively and resulted in an even stronger victory. This element of ambiguity turns playing a game into an art. As with learning to play the guitar or learning to paint, you improve through exploration, and also through absorption of *guidelines.* In a painting class you learn guidelines for using color, composing a painting, using texture, mixing paints, and even holding your brush. But any good teacher will also tell you that these principles are not *rules*, but guidelines. There is no one solution to the problem of how to paint well. Artists can violate all kinds of guidelines and still become successful and beloved—history is filled with such stories.

This is exactly how it is if you want to become better at a game. There are guidelines for good play, such as generally not getting close to a Heavy Weapons Guy in *Team Fortress 2*—his damage output is so high from close up that you're generally dead within a second. However, there are exceptions to this: a notable example is that if you're a quick-moving, double-jumping Scout, you can sometimes bewilder a Heavy by double jumping over his head and around him, like a fly. This is only applicable in certain situations, though, and is very much dependent on a number of variables, such as where you are in a given level, what other classes are around, how much health you have, etc.

This is just one of thousands of examples of guidelines and exceptional subguidelines. There are subguidelines that go below that, and *all* of them can be ignored with great success in the right place at the right time. In this way games reward not just study, but also ingenuity and innovation. A truly great player knows not only the guidelines, but also when to throw them out the window and try something bizarre. A deep game allows this.

Contests are starkly different from games because they lack these kinds of decisions. Could you argue that a push-up contest does include some decision making? Of course. It's impossible for a human being to be alive and not make *some* decisions. For instance, you could decide to hold your breath on every third push-up, or inhale on each push-up. Or you could decide to think about a certain montage from the film *Rocky*, or instead try to clear your mind. These are all examples of the kinds of decisions that can be made during contests, and pure puzzles involve similar types of decisions. They are markedly different from the decisions players make during games, however, in that they are not directly relevant to your interaction with the system—in other words, they are not *endogenously* meaningful. I'll go more into detail about what this means later in Chapter 1, but for now it suffices to say that decisions in games have effects inside the system—a butterfly effect that affects the whole web for the rest of the match.

Thinking about Games

In the next section I'll get into what I mean by a meaningful decision. But before I do, I need to go into a little detail on the difference between the *abstract* and the *literal* in games (and other systems). The word *meaning* can itself have a lot of different connotations, but I define it in a very specific way.

Some may think that a decision is meaningful if its implications cause the user to think about some deep, insightful, or personal issue not directly related to the game itself. These types of meaning are what I'd call *thematic* meanings, and they don't reflect at all the way that I'm using the word. In order to explain this distinction, I'll have to talk about some differences between the abstract and the thematic.

If you were to replace the artwork in *Super Mario Brothers* with nothing but colored squares, you would turn the game into one that is more abstract. Games that are abstract use *representative* art or mechanisms (Figure 5). It's easier to tell when a game's artwork is abstract than

Jump (Literal)   **Jump (Abstract)**

Figure 5. Literal versus abstract game art.

when its gameplay mechanisms are. For instance, you could represent a soldier on a map with a helmet or sword icon, which would be a clear example of visual representation. It's harder to give a clear example of gameplay abstraction, though, because game rules are inherently abstract (Figure 6).

A very clear example of an abstract or representative mechanism is the classic health bar of so many digital games. Usually in-game avatars have numeric expressions representing their current physical status: high numbers represent health, lower ones mean injuries, and zero usually represents death. There are many reasons that game designers render some things abstractly or representatively and other things literally or thematically, and I'll get into these more in Chapter 4. For now, it's sufficient to say that a great emphasis is put on the thematic elements of video games, which is causing us to actually miss the point of games: developing a strong set of rules.

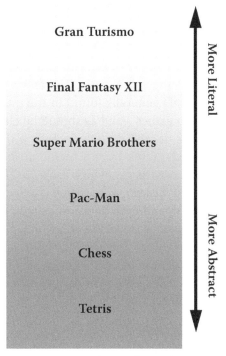

Figure 6. A ranking of popular games from the literal to the abstract.

The Meaningful Decision

Let's say you're playing a highly thematic game—perhaps a game like *Final Fantasy*—and a prominent character dies in a cutscene (and I mean *really* dies, not the type of death that can be fixed with a "phoenix down"!). As that character dies you may think of a loved one who has passed away, so you walk away thinking that *the game* has personal meaning to you. But strip away that theme and look just at the mechanisms behind it. If it's *Final Fantasy*, your party—which is essentially a group of integers and Boolean flags that contain various resources, utilizes special actions, and lets you do a certain amount of damage in combat—just lost one group of resources. Obviously, if looked at in this abstract state—the *true* game-state—there's no reason that the game would be making a person think anything about a dead loved one, and so it is clear that it's the game's theme and not the system itself that is conveying the sense of personal meaning. We can take this example a step further by attaching the image of a dead character to just about anything—a poster, a video, even a lunchbox—where it could very well evoke the same reaction that it did in the video game. Therefore, it's clear that these kinds of meanings aren't generally found in the mechanisms but rather in the themes of video games—and as should be clear by now, my proposed definition of *game* doesn't directly include a thematic layer. This isn't to say that games shouldn't have a theme, just that a theme should only support game mechanisms, which are *what the game actually is*.

So, a *meaningful decision* is a decision that has effects inside the system. A meaningful decision usually has a rippling effect in a game, and not all effects can be known, which makes the correct choice ambiguous. Modern video games are rife with decisions that are *not* meaningful and are merely thematic, meaning they have very little effect on the system itself. These decisions are almost always false choices, and I'll get into that more in Chapter 2.

Are Games Art?

There are many definitions of art—if you ask a hundred people you'll surely get a hundred different answers. Personally, my definition for art is *the product of human creativity*, and this definition is quite close to most relevant dictionary definitions. Merriam-Webster's fourth definition of art is shown, in part, below.

> Art: the conscious use of skill and creative imagination.

At the time of writing, the first line of the relevant Wikipedia page defines art in a similar way.

> Art is the product or process of deliberately arranging items (often with symbolic significance) in a way that influences and affects one or more of the senses, emotions, and intellect.

Trying to define art more precisely does very little except to exclude some kinds of works arbitrarily, so I don't think it's productive. (Although many people seem to get something out of it, and I doubt that they'll stop anytime soon.) The question of whether games are art is not productive because the answer depends entirely on one's definition of art. By my definition (and that of most dictionaries), however, games are absolutely works of art.

In most cases, the question implied in the "are games art" debate is more sinister and boils down to this: are games a legitimate thing for us to enjoy? Do we have the approval of some elite class for these things we love? At the end of the day, it's not important whether you, I, or anyone else thinks games are art. The value of games to human beings is undeniable. People can say that the film *Rocky* doesn't meet their definition of art, but that doesn't affect its impact and value to those of us who enjoy it.

Games: The Finer Interactive System

To be clear, I am not about to say that games are better than puzzles or contests; each has their own kind of value to people. What I am saying is that games are a much more fragile and unstable thing. Games, if not carefully built and maintained, break down into contests, puzzles, or even basic interactive systems.

The ambiguous-decision property of games is surprisingly elusive. In creating a game, you have to create a system deep enough that a human mind (which is a very powerful thing) cannot master it. Mastery is a bad thing for a game—even if it takes ten years to attain it—so mastery needs to be *unattainable*. On the other hand, you don't want the system to be arbitrary or just pure noise. There has to be a reasonable path towards mastery that players can take. If a game feels as though not only mastery but even mere competence is impossible, then players will almost surely abandon the journey of learning it.

Games must dance upon the threshold of the known and the unknown. They must live at the border between what we can understand, and what we cannot. This border is very narrow. On one side are puzzles and contests. If the game is solved, it's effectively a completed puzzle, and if an element of strength or dexterity is required, then it may qualify as a contest. For instance, if someone "solved" boxing (not going to happen), winning may simply become a matter of who can deliver the solved moves faster and harder. And if something becomes a raw measurement of physical ability, then it is a contest, not a game, since no decisions are left to be made.

On the other side of the border there are simply games that are too difficult to process—games that people *can't* pursue mastery in, and that they can't even comprehend. If players can't even grasp the mechanisms of a game, then their decisions will be totally random and meaningless and they cannot pursue mastery. Note that I didn't say that games like these aren't games; however, I might call them bad games.

A Fragile Illusion

Like all things usually referred to as art—works produced using creativity—games are an illusion, a trick that we play on ourselves. Just as a painted portrait is actually a thousand little dabs of paint and not the face that we see in our mind's eye, a game is actually just a solvable puzzle. In the same way that one errant stroke can mean the difference between a believable landscape and a breaking of the illusion, one errant rule or imbalanced element can break the illusion of ambiguity. Tic-tac-toe is a game to children, but for adults it is a simple, solved puzzle. Perhaps to some far greater alien intelligence, chess would appear the same way as tic-tac-toe appears to adults.

In this sense, whether a system is a game or not comes down to a certain specific type of subjectivity. However, it's the kind of subjectivity we can easily nail down and set aside by simply asking the question, have you solved it? I'll get more into *solved games* in Chapter 2.

Game Playing Itself Is an Art

Most of us agree that writing music, graphic design, and architecture are examples of art forms. What distinguishes them from other activities that are not art forms is the fact that they require creativity. The reason that they require creativity is because they are trying to achieve a goal (often a message or a feeling) within a framework (the human mind) in which the optimal situation is not known. Sounds familiar, right? When people design games, they are designing new forms of art. Game players are artists too, using inspired creativity and ingenuity to come up with new strategies and gambits that hopefully push the understood limits of that system in a new direction.

In the arts we don't have optimal, absolute answers. Instead, we have guidelines—best practices that tend to be helpful. For instance, music theory tells us that generally we don't want voices to cross (except when we do) and that we usually don't want a minor third and a major third in the same chord (again, except when we do). Photographers generally want to divide compositions into thirds (in accordance with the golden ratio), but again, there will always be times when they don't want to do that. This is why these principles are guidelines and not rules.

Game players have the same sorts of things going through their minds. If you look up gameplay strategies for games such as soccer, chess, Go, and poker, you won't find anything like "on your third turn, you must always do this." Guidelines are instead conditional and use terms like "usually," "you may want to," and "depending on the situation." Playing a great game well is a matter of taking guidelines under consideration but also being free enough to shed them when you see a unique opportunity crop up. Games that have been solved do not allow for this, and by my definition no longer qualify as games for this reason (perhaps you could call them "dead games," since they were originally intended to be games). Playing them is no longer an art because there is no longer any creativity involved. The element of creativity is what makes games so interesting, and also what makes them so difficult to create. They must restore the feeling of creative play (which exists in basic interactive systems, but is lost in contests and puzzles) within a highly structured type of system. Game designers are doing more than just creating art: they're creating arts.

The Value of Games

The unique combination of problem solving, competition, and the afore-mentioned ambiguous decisions in games make them unique types of machines that have great value to a human intelligence. If you were to ask what the value of games are, most people would say that games are fun. However, this answer is not precise enough for a game designer— and further, it's not even always true. The primary and direct value that games have for us is that they teach us how to learn. They provide an environment for us to focus on increasing a specific skill or set of skills. They teach us to formulate tactics, to second-guess our thinking, and to commit to a strategy. Quite simply, they allow us to train ourselves as *thinkers*.

It's true that the byproduct of this training can usually be called *fun*. What's actually going on, though, is that humans (being highly cerebral and curious creatures) have a natural biological hunger for further un-derstanding the world around us. When we experience a feeling of self-improvement, or a feeling of having learned something new, our brains reward us by releasing endorphins. The pursuit of mastery is exciting for us; this is one of the things that makes human beings very special. And it's also one of the things that makes games in particular so important and valuable to us. Our studies of humans and other primates make it very clear that our curiosity—our need to *understand*—is a biologically important element that's on par with our need for companionship.

Games have great philosophical and social implications for us; in a way, they help us to understand who we are on a very basic level. In Frank Lantz's 2011 Game Developers Conference talk, "Life and Death and Middle Pair: Go, Poker, and the Sublime," he talked about the game Go—an utterly abstract game—and all of the meaning he sees in it. He described a Go match in progress as a way of seeing two minds entangled in intense battle—all of the testing, pushing, pulling, responding, and reactions literally can be seen, forming a complete web that illustrates a human discussion. Which gets me to my final point about games and their value: games are beautiful. Anyone who has an appreciation for nature will delight at the beauty and perfection of a game that is brilliantly designed. In nature, systems of rules in which agents compete by making ambiguous decisions spring up all the time, and all around us. In this sense, the game designer is trying to simulate a nature that never was.

Misconceptions about Games

In today's culture, the word *game* has a lot of negative and positive values associated with it that are unfair and incorrect. Before we go on, I think it's important to address these head on.

- *Games are for children.* As of this writing, the average age of a game player is 37.[2] Furthermore, when you take games as a whole—as opposed to just digital games—it becomes clear that games are not just for children. For instance, checkers and contract bridge usually are associated with the elderly, and competitive games like Go or sports like football are clearly of great interest to people of all ages.
- *If you're saying x is not a game, that's an attack on x!* I have encountered many people who got upset when I said one of their favorite video games was not a game. *World of Warcraft* has more in common with a theme park than it does a game, and *Garry's Mod* has more in common with Legos or a sandbox than it does a game. This does not at all speak to the quality of those things, however. Is it an insult to theme parks to say that a theme park is not a game? Is it an insult to a sandbox to say that it is not a game?
- *It's possible that the current cultural rock-star status that games have achieved has attached a certain silly cultural value to the word* game. We have to remember that *game* does not mean *good*. The worst game you ever played is still a game, and your favorite thing can be something other than a game and still be just as legitimate as if it were one. We simply need to be consistent with our words.
- *Fun is a fundamental part of games.* As I've explained, fun is not a building block of games but instead is a byproduct of games in action. Much like the art issue addressed earlier, it also depends on how you define the word *fun*. A game can be a completely mis-

---

[2] Entertainment Software Association (ESA) 2011 report (http://www.theesa.com/facts/index.asp).

erable experience that few would define as fun and still be a game. It could still be a *good* game, as seen through my lens. To use an example horrible enough to illustrate my point: a knife fight to the death would probably not be any fun at all, but it is still absolutely a game (and there are people who love, and study, the art of knife fighting). A real life example might be fights in the Colosseum—which I'm sure were not much fun for participants, but which were still deep, rich, and interesting games. But we need not even look to those kinds of examples to find games that are not fun. What about games that are simply bad? We've all played at least a few games that we personally did not find fun (many of us might even say that most games are not fun), but that doesn't mean they aren't games. Personally, I design my games with fun as a desired outcome, but it's not a fundamental part of games. One does not "add fun" to a game.

Games Can Occur Naturally

There is nothing about games that says that they must have been designed intentionally by a human being to be games. Think of it—many people's workplaces function as games: there are goals (solutions), there is competition, and they often require that the "players" make meaningful, ambiguous decisions. Of course, if your job is on an assembly line it may be simply a contest. Fistfights also qualify as games, since there is a system of rules in place, in the sense that it's taking place in the physical universe. There are also goals (solutions), and there are most certainly very interesting decisions to make.

However, it is often the case that naturally occurring games usually don't make great games in a pure sense. The problem with naturally occurring games is that they are messy—there are many "meta" elements that can get in the way of the game mechanisms. In the case of a street fight, for instance, some third party may attack you suddenly, or the other "player" may pull out a gun, or you may stop the fight early for fear of being arrested or injured. This is the reason that designed games tend to be isolated, somewhat abstracted, self-contained systems: so that players can focus fully and completely on the mechanisms themselves, which allows them to reach their full potential in the game. Very few people have been able to really explore the "game" of real-life street fighting because of the legal and physical risks involved in such a game.

As I mentioned earlier, when we create a game we are trying to mimic a nature that never existed. We must tap into the same concepts of asymmetrical balance (but not necessarily asymmetrical forces, which are covered in Chapter 2) that surround us in naturally evolved systems, and we must harness them for our own purposes.

Video Games and the Value of Words

As I mentioned earlier, it is my opinion that the term *video game* has caused an incredible number of problems for the players, creators, and marketers of digital games. This label has made things harder for everyone involved in the world of digital games. The term is far too broad and encompasses many things that are very different from each other. It is difficult to judge interactive systems by the same yardstick if one is a puzzle, one is a contest, and another is a game.

Simulators and puzzles are not genres of games, but the use of the term *video game* sends that message, with significant ramifications. Some of the negative effects of this term follow.

- ◆ *For both consumers and marketing people.* The product being sold is undefined. There are defined genres—such as first-person shooters (FPS), real-time strategy games, and role-playing games (RPG) (and I should note that the genre is often not listed either on a game's box or on its promotional materials)—but these categories can be vague and not provide a true picture of what the software actually is. For example, a game could be called a platformer, but that doesn't tell you whether the software is a game (such as *Spelunky*) or a puzzle (such as *Braid*).
- ◆ *For critics.* People using the same yardstick to judge a dry-flight simulator on one hand and *Super Mario Galaxy* and *Street Fighter* on the other are almost certainly going to have difficulty. Different interactive systems are trying to achieve different things, and few critics are expert enough in all of them to provide useful insight for any of them. This is a major reason why video-game criticism is generally little more than a summary of what the game is and whether or not the reviewer enjoyed it.
- ◆ *For creators.* Some design decisions make a lot of sense for a simulator but almost no sense for a game. For example, having to worry about your fuel in a dogfighting simulator makes sense, but you might want to disregard that element for a dogfighting game so that players can focus on the aspect of dogfighting related to making interesting decisions. (There will be more on this in Chapter 4.)

The worst part of the term *video game* is that we lose the ability to identify pure games that happen to be played with a computer, such as

Tetris or *StarCraft*. For this reason, I'd like us to be a bit more specific when we talk about interactive entertainment software. I call games that are played on a computer *digital games*. I might call something like *SimCity* a simulator, *Portal* a puzzle, etc. In the future we'll probably have to come up with some new words to describe various things. I'm an advocate of the newly born term *app* that has come to describe digital interactive applications on smartphones in the last few years. It is a much less harmful and more accurate term than video game.

The important thing to take away from this chapter is that a simulator is not a game genre, and that we need to use language that's more precise. It's important that we use language in a productive way, especially for those of us who are serious about games and pushing them to the next level.

Exploration

Some games are said to have an *exploration* gameplay mechanism. Usually this means that there's some kind of unrevealed overmap, shrouded in a fog of war or hidden behind secret doors. This is what I might call literal, or thematic, exploration—but if you understand one thing about games, it should be that they're actually *all* about exploration.

Games are inherently an exploration, or a discovery, of a *possibility space*. Playing a game is testing the limits of a new reality. When players win, they know in the back of their minds that they could have done even more; when they fail, they imagine other routes or actions they could have taken to succeed. Games are microcosms of life in this way: we are plopped down into this interesting world and we comb through the information presented, trying to make sense of it. We search tirelessly for the answer—the solution—but we never find it. It's a constant, strange, mysterious, exciting, lateral brainstorm that we wish would never end.

One of my favorite BoardGameGeek users, J. C. "clearclaw" Lawrence, had this to say about the board game *Age of Steam*, and I think it captures a lot about why games are special:

> After playing, your mind quivers, not in shock or burnout, but in exactly the same way your legs will after pounding your way up a steep hill: in the riotous enjoyment of being alive and working hard and knowing that next time, next time, you can do better.

2

On Game Design

D esign is essentially a synonym for plan. When we design a game, we not only have to plan what kinds of actions will be possible in a game, but also all of the types of *inter*actions that could take place. Game design is all about planning. In this chapter, I'll be giving a detailed walk-through of what I believe to be the best approach to creating games that will stand the test of time. The best way to start is to filter out any *bad* reasons we may have for wanting to make games.

The Medium and the Message

The first thing to ask when you're about to embark on the journey of game design is, am I sure that a game is what I want to make? Because of the almost limitless technology available to a digital game designer, it's tempting to believe that games are a good place to express just about anything. While it's true that you can express anything in a digital interactive application, a game is a specific thing. A game is a system of rules in which agents compete by making endogenously meaningful decisions.

This sort of a system is very good at expressing abstract concepts such as spatial relationships—as in, this object is above that object—and numeric expressions—or, agent A has more of resource X than agent B.

These kinds of relationships also extend into larger themes such as territory control, prediction, adaptation, risk management, and many more. Often, the real themes of deep games (such as Go) are difficult to put into words. As noted in Chapter 1, game designer Frank Lantz has described a game of Go as a complex visual representation of the intertwined thought patterns of two players. This is the sort of theme that you can actually abstract from game mechanisms.

Can you also express a literal theme, such as love of a father, in a game? The most obvious way to do this would be to add nongame material such as cutscenes or dialogue (i.e., cinema or prose) in order to express your theme. Although our broad usage of the word *game* to refer to digital entertainment software may lead us to say that the game is expressing the theme, in reality the *game* parts of the game cannot do this.

In more recent years, some developers have taken to the task of expressing a literal idea through the use of an abstract system. Rod Humble's *The Marriage* is one such example. In *The Marriage*, the player loosely controls two squares on a single screen. You have various ways of making the squares grow and shrink, and overall the application is meant to say something about marriage itself. But is *The Marriage* a game? I would say definitely not, because it lacks a goal—it's not a contest between different agents. Several examples of so-called *art games* (a term that I personally find offensive) make similar attempts to express ideas. However, in every case it's clear that a game is not making a home for literal expressions. If your goal is to express a literal idea, there are almost certainly better media to do so.

Questions to Ask

Here are a few questions that you should ask yourself before you start to design a game. If your answer is yes to any of these questions, you should consider another medium. Remember: games aren't better than other media. In our culture and era, it can be easy to fall prey to a misguided desire to make a game when you'd be better off making something else.

- *Is your goal to tell a story?* Consider a linear, temporal medium such as prose, cinema, or comics. I'll get into this more in the next section.
- *Is your goal to feature a character?* Again, stories tend to be the best way to reveal who a character really is by showing the decisions he or she makes. In a game, the player makes the decisions, not the character.
- *Is your goal to feature a physical object, image, or setting?* I have heard many novice game designers describe their ideas for games

that are little more than an idea about a magic sword, or a post-apocalyptic desert world, or some such thing. What should be obvious is that these aren't ideas for games. All of these are great subjects for portraiture, since they are visual or verbal descriptions of persons or things. Consider a painting, a photograph, or video art instead.

Really, the question to ask first is: will interactivity help me do what I want to do? Further: will a *game system*, with its goals, its competition, and its player interaction be helpful? The answer may very well be no, and you'll be much better off if you start in the right medium.

Games and Story

First, I should make clear what I mean by *story*. I often see people using this word to describe an *emergent* story; a story that unfolds as a natural process of any game. This is not the kind of story that I am talking about here—obviously, any game (or really, any activity) will yield an emergent story. What I'm talking about is a prewritten narrative, as is seen in games like *Final Fantasy VII* or *Half-Life*.

Since digital-game technology allows for stories in games, more and more games include them. Some video games that are considered the greatest of all time not only *include* story but are actually *based* on story. Games like *The Legend of Zelda: The Ocarina of Time, Final Fantasy VII, and Metal Gear Solid* set the standard for modern video games. Very few in the game-development world are willing to challenge these sacred games, which I think limits the new games we create to only being *as good* as them. The real question is, does the presupposition that games should have a story help or hurt digital games?

Let's look again quickly at our definition for a game.

> Game: a system of rules in which agents compete by making ambiguous decisions.

The dictionary defines story as "an account of imaginary or real people and events told for entertainment." This definition is fine, but for our purposes it may be more useful to define *story* using the language below.

> Story: a telling of a sequence of events.

It's important to note that a story is essentially a list. An account cannot have two possible first events, since only one thing can happen first. A story is sequential: this, then this, then this. We therefore can draw the experience of *story* in a straight line with nodes representing

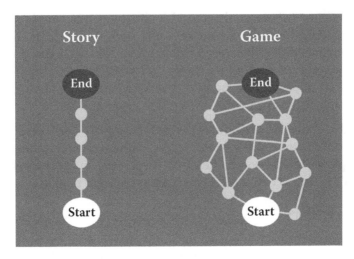

Figure 7. A rough representation of the shape of a story versus that of a game.

various events. I need to make clear that I'm by no means saying that stories are simpler than games. Both stories and games are complex "machines" that actually have to function, each in their own way. Good stories have many threads that interweave with each other in a graceful and beautiful way. In terms of what the user experiences, they are linear lists of events (Figure 7, left).

Games, however, do not consist of linear lists of events (Figure 7, right). The experience is more like a constantly evolving and emerging web, since as players go through them, the nodes and connections (the possibilities and choices) are changing. It's not always clear to players how nodes are connected—in fact, getting better at a game is a process of getting better at predicting the future structure of the web. When we can completely map out the entire web of a game (as we can do with any story we've seen before), the game actually is "solved" and becomes useless to us (think tic-tac-toe, in which most adults know with certainty the optimal move in any situation). So those who are interested in making a story-based game essentially are left with the three options below.

- *Cutscenes.* The most common way to create story-based games is to use cutscenes. With this method the application essentially bounces back and forth between a movie and allowing the user to play the game parts. It has become more clear to most developers that this method is a somewhat sloppy solution and players will probably grow more and more irritated by it as time goes on, since having a play experience interrupted is frustrating. *Metal*

Gear Solid, *Gears of War*, and *The Legend of Zelda: Ocarina of Time* all rely on cutscenes.

♦ *Allowing the story to trump interactivity.* When the story trumps interactivity it's sometimes referred to as "being on rails." Apps that exhibit this model of story typically give players very little choice over how the game goes. Often they're required to simply move down a linear corridor for the entire length of the game, sometimes with small bits of "gaming" thrown in to keep it interesting. *Dragon's Lair*, *Final Fantasy XIII*, and *Half-Life 2* all use this model.

♦ *Allowing interactivity to trump story.* Games that allow interactivity to trump story are usually the best ones, but the roles and quality of the stories are greatly diminished. Most games in this category would be just as good without any story at all, and it's often fair to say that the story is somewhat tacked on. *Super Mario Brothers*, *Katamari Damacy*, and *100 Rogues* all put interactivity above story.

The High-Tech Solution that Will Never Be

Some people think that one way to integrate stories and games better would be to have the game system regenerate the rest of the story in response to the decisions a player makes. That is, all of the character arcs would respond to each decision, in real time, regenerating themselves into a new complete story that would be satisfying and say something powerful and resonant about life. To me, this idea is ridiculous at best and impossible at worst. If you're one of these people, allow me to explain why this idea will not be useful within our lifetimes, and why it will probably *never* be useful: *writing good stories is very difficult.* Experienced writers spend weeks or even months on a single scene. Changing one decision in an otherwise great story can turn it into an incoherent soup of nonsense. Each time a character makes a decision in a story is a chance to break the story—stories are fragile machines.

So if a computer is making real-time decisions about how a story should continue in response to a player's decision, it has to be *far* more intelligent than the greatest human writer who has ever lived! That alone puts this idea well into the future. But even if we did have that fantastically smart computer, we still have another problem—there's bound to be some decision combinations that simply *will not* add up to a coherent story (let alone a good one).

Another solution, of course, is just to accept that most stories in games will be bad or mediocre. I'd like to believe that none of us want to shoot for mediocre, though.

Understanding Design

> Design: a plan or drawing made of an object before it is created that illustrates its form, function, or mechanics.

One of the most important aspects of learning game design is coming to a better understanding of the essence of design itself. For many of us, *design* can be one of those words that we use but never take the time to fully understand. Design is important, and detailed standards for what constitutes good design are paramount for all designers, whether they're designing cell phones, dresses, houses, or digital games. Good design is many things, and there isn't complete agreement on what those things are, but here are four characteristics to consider.

- *Useful.* A good design solves some problem—whether that problem was one we knew about or not. We often find ourselves wondering how we lived for so long without Well-Designed Product X only a short time after acquiring it.
- *Beautiful.* A good design has a certain kind of beauty to it. This doesn't have to be a visual beauty; it can be an abstract sort of beauty like that of the rules of *Tetris*. Game rulesets are often very beautiful in the sense that they fit together very well and at the same time unfold into incredible networks of possibilities.
- *Easy to use and learn.* A good design almost "uses" itself. The user doesn't struggle and hit brick walls; a great design provides a smooth experience from start to finish. The "It Just Works" advertising campaign Apple used to sell Macintosh computers was an attempt to sell people on the idea that Macs exhibited this property of great design.
- *Efficient.* A good design does a lot with a little. Great painters know how to express their vision in as few strokes as possible. Great poets know how to say what they want in as few words as possible.

I think that there is one word, though, that sums up all of the above: *elegance.* In short, design is doing something well, which doesn't seem all that helpful by itself. But we should all try to take apart the essence of design and find out for ourselves what makes something "done well" in terms that apply to *all* of the arts. We should all strive to formulate clear guidelines for what will make our work valuable to human beings.

Minimalism

There was a movement (associated largely with the 1960s and early 1970s) called minimalism that still lives on today. Its core tenet was that art—whether painting, clothing, architecture, theater, or anything else—should be no more complex than it needed to be. I feel that this is true for all art, and that minimalism is almost a synonym for grace or elegance. Good design is always minimalist: even if you're painting a picture of a crowd, you should paint only as many strokes as are needed to express that idea. Everything should contain as few words, strokes, gestures, or rules as possible—that *is* design.

Keep in mind that minimalism doesn't mean a lack of ornamentation or complexity. A game's rule book can be as thick as a phone book, as long as it contains the fewest number of rules that make it possible for you to express what's important about your game.

Although I believe the principle of minimalism applies to all arts, I think that it may be even more applicable to games. As I mentioned in Chapter 1, playing a game is an art form. Game designers are providing a way for players to express themselves, and in the same way a designer of a guitar needs to consider ergonomics or the designer of a theater needs to consider acoustics, game designers need to make sure that there are as few obstacles as possible between players and their self expression. It's hard enough trying to express yourself—we don't need to make it any harder.

Core Mechanism

When you start designing your game, you should begin with a *core mechanism*. Often, you'll want this to be something rather simple—almost abstract. Examples of core mechanisms include *jumping, bidding, deduction, pushing,* or *aiming.*

A game design should always start with a core mechanism, and from there you can figure out how many interesting interactions will surround that mechanism. Ideally, every single thing that's inside the game should be in direct support of the core mechanism—and anything that has nothing to do with the core mechanism can probably be removed from the game. Keep in mind that there are plenty of games whose core mechanism isn't easily labeled, particularly some of the more interesting abstract games. For instance, what's the core mechanism of *Tetris*? Is it *placing*? *Rotating*? *Filling*? I can't really think of a specific word that defines it, and it would even be hard to describe using a whole sentence. Regardless, it's clear what the core mechanism of *Tetris* is and how it ties into all the supporting mechanisms. Conversely, some games either have no core mechanism or supporting mechanisms that have almost nothing to do with the core mechanism. For example, what's the core mechanism of the popular board game *Cranium—rolling* and *moving*? What does that have to do with answering trivia questions?

Beware the Excesses of the Digital Medium

Human beings have a bit of a bad pattern with technology: when we first unlock a new technology, we never ask whether we should use it. Instead, we tend to use it indiscriminately at first. It's only afterward, when the technology has become established, that we become more selective in its use. We need to reach that place with respect to digital games. Computers can handle incredible complexity, and most developers have little concept of restraint when it comes to using the tools at their disposal. This is why I think designers who want to reach a higher level of computer-game design should use board games as their inspiration—they are always limited by their physical requirements, and they tend to have much more sensible designs for this reason. I'll talk more about them in Chapter 5.

Let's Add Some Fun!

What is fun? Is fun just simple joy, enjoyment, or pleasure? Or does it mean that you're engaged? Personally, I feel that games should be enjoyed, but they don't have to be enjoyed to be great games. The bottom line about fun is that there is little agreement about what it means, so for now I'll be using it in a very broad sense that includes enjoyment, engagement, and fulfillment.

When I present my philosophy about games the response I sometimes get is, "all that matters to *me* is that the game is fun." As I stated in my introduction, everything in this book is written for the sole purpose of creating machines—games—that are as much fun as they can be (using my broad definition of fun). If it weren't for that goal, no designer would do any of the thinking, researching, testing, and everything else we do for the sake of making better games. In short: *fun is the name of the game.*

The issue, of course, is that we cannot simply inject *fun* directly into a game. Fun is actually a byproduct of a great game design, which is why I think defining our terms is more useful than talking in general about fun. Questions about whether something is an interesting decision or whether element A is balanced against element B provide us with more objective answers. (You should be clear on this: someone who says, "I don't know, I thought it was fun" is not really saying anything. In fact, this kind of statement is often an attempt (conscious or not) to shut down the conversation when the speaker isn't contributing anything objective to the conversation. When we talk about games, we must use characteristics that are objective and quantifiable to illustrate our points.)

For instance, let's say I want to explain why I think *Puerto Rico* is a good game. For those who aren't familiar with the game: players take turns selecting roles such as Captain, Mayor, Trader, and so forth. Each role has a different power, but all players perform each role's action. The player who selected a given role gets a small bonus in effectiveness. Instead of simply saying, "*Puerto Rico* is fun," I would say that its multi-layered turns system (turns within a turn, with each player taking the action of the selected role) means that the system is very flexible and the possibility space is very deep. I would also add that all of the strategies I've encountered in the game are well-balanced, and that the game continues to surprise my entire gaming group even after hundreds of plays. These are all hard, factual observations about the game, which form a constructive argument for *how Puerto Rico* is fun.

Engaging, tense, interesting—all of these are often much better descriptions of what many games have to offer people. Tension in particular is a somewhat fundamental part of games, and is a direct result of an interesting, difficult, (and most important) *ambiguous* decision.

Establishing Standards

In a nutshell, the purpose of this book is to give us the tools that we need to judge a game objectively. Of course, with any of the arts not everyone will agree on what makes something good. If you look at other media, however, you'll see that there are some things that are generally regarded as fundamental to the medium. These are guidelines, not rules, but they are extremely helpful and we need to establish what they are sooner rather than later.

Nonlinearity

Games are *inherently* nonlinear. We observed in Chapter 1 that the non-linearity of games is the biggest barrier between game and story. Because games are interactive systems, they are necessarily a web of "possibility nodes," as opposed to a linear list of events. What's interesting is that this characteristic affects every stage of game development, especially design. The process of game design itself is a nonlinear act: when describing your system, you are forced to talk about mechanisms that you haven't yet explained. You'll notice this when reading a manual for a complex board game, such as *Agricola* or *Battlestar Galactica*.

For instance, the "Game Turn" section comes early in the *Battlestar Galactica* rule book. This section lists the steps of a player's turn, one of which is the Activate Cylon Ships step described below.

> Activate Cylon Ships (if necessary): if any are in play, Cylon ships are
> activated according to the Crisis Card drawn...

At this point, obviously, we have no idea what Cylon ships are, how they "activate," what a Crisis Card is, or how a Crisis Card might interact with a Cylon ship—and yet here it is, something telling us exactly that. But the truth is there's really no other way. The writer realizes that readers don't have any of this information yet, and so continues

...(see Activate Cylon Ships step on page 11).

This cross-reference lets readers know that it's not their fault that they don't know what the author is talking about. The fact that games are nonlinear means that their manuals are also nonlinear, which makes them difficult to read.

Nonlinearity and Game Design

Why am I talking about nonlinearity? Because this characteristic of games is something designers need not only to realize, but also embrace. Games are inherently nonlinear, so every step of the way the process of game design and game development will be nonlinear as well. One of the first steps we take in designing a game, for example, is writing a game design document. Of course, a game design document is (usually) a linear text document, and so we're back to the same issue we had with the manual. If the first paragraph describes the uses of the *jump* action in *Super Mario Brothers*, and one possible use is "to release power-ups from special blocks," readers will be totally confused about what that use is.

For this reason I advocate using a nonlinear (web-type) format for writing game design documents. Personally, I like to use online wikis, which allow for several different pages all linking to and from each other. Wikis also save all of your past edits and revisions, which can be useful in the ever-changing game design process. There are many sites where you can create free wikis online, and they're great tools in any developer's arsenal. There are other options, too: you could create a web pattern physically by laying out Post-it notes on a wall or index cards on a floor. Either of these are a really great way to see relationships that a text document will never show you.

The Designer Is the Bad Guy

The nonlinear nature of games makes it incredibly hard to predict all of the possible ins and outs, possibilities, and offsets of a game system. This means that as a game designer, you're going to be wrong—a *lot*. If you're not a person who's comfortable with admitting you're wrong about something, then you either shouldn't be a game designer, or you should be a lone-wolf game designer who can create a game with no other help.

I'd say that by the time a good quality game is finished, there isn't a single design aspect of the final game that the designer wasn't wrong about at some point in development.

What this means is you're probably going to make those you're working with mad. Programmers will spend hours working on some special spell or world-generation algorithm, only for you to tell them a week later that these things have been cut. Or an artist will spend hours meticulously placing pixels in a character sprite for a character that has to be eliminated due to a balance issue.

With good project planning that includes a long, thorough design process up front and a lot of prototyping you can minimize this type of thing—and you absolutely should—but you will never eliminate the problem. Tell your team you're sorry in advance.

Continuous and Discrete Space

Every game designer needs to understand the distinction between a *continuous* and a *discrete* possibility space. A strong understanding of these concepts will allow you to make better judgments about your own game designs, and also are important for understanding some of the terminology in the rest of the book.

It's easiest to speak first about discrete spaces in games, because they are the more abstract (and perhaps more artificial) of the two types of space. Discrete space is a space in a game (usually a tiled or divided area) wherein the entirety of the space means the same thing to the game (Figure 8, left). The basket in basketball is an example of a discrete space: putting the ball through that ring—a discrete space—awards that team two points. It doesn't matter if the ball hits the back of the ring, the front of the ring, or directly in the center (making that nice *whoosh* sound!): for the purpose of getting points, all that matters is that the ball passes somewhere through that ring. Chess and most other abstract board games are composed entirely of discrete spaces; indeed, a grid is precisely the dividing up of a playing field into discrete spaces.

On the other hand, continuous space is found on a soccer field or in a real-time digital game such as *Quake* (Figure 8, right). When you fire your rocket launcher at opponents in *Quake* it's wise to fire at their feet, so that even if you miss, your opponents will take splash damage (damage from the explosion of the rocket). In this case each tiny pixel (or inch, or whatever in-engine unit of distance is used in *Quake*) matters. If your rocket explodes just one inch from an opponent, it will deliver more damage than if the rocket explodes two or three inches away (and after all, damage is the object of the game, at least in a head-to-head deathmatch situation).

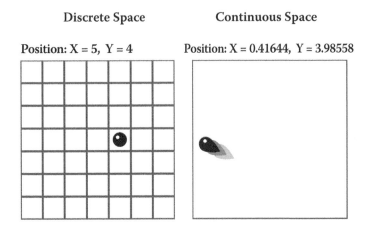

Figure 8. Examples of discrete and continuous spaces.

Of course, inches themselves are discrete units of measurement that we use to describe and record length, but in the *Quake* example the distance to the explosion is effectively continuous (even though the in-game system probably has some kind of grid if you look at a small enough level of detail). Because games always have to have an end condition, which is necessarily discrete, no game is entirely continuous. The goal in soccer, for instance, is a discrete space. Any game with continuous space is going to have some discrete space at the end, if nowhere else, to provide the condition that ends the game. An example of this would be the sport of fencing, in which the first competitor to fall prey to three *touches*—getting hit by the opponent's weapon—loses the match. A touch is defined differently depending on the type of fencing, but usually there is a discrete area on the body that is considered valid for touches. Judges watch carefully to discern whether or not this discrete area has been hit by the opponent's weapon.

Continuous and Discrete Time

Time itself, which is normally continuous, can also be divided into discrete segments. Any game that is turn-based divides its game time into discrete spaces that we call turns. It doesn't matter if you take two seconds or two minutes to take your turn in chess; both have the same meaning to the game. Alternatively, in a game based on continuous time—also known as real time—such as *Super Mario Brothers*, it absolutely *does* matter precisely when you decide to do something. Because monsters are constantly moving onscreen, jumping right now results in a very different outcome than jumping even an instant from now. Instants matter in a real-time game, but only turns matter in a turn-based game.

Input Resolution

Each game that is designed has its own *resolution of input*. This is essentially the size of the data chunk that can be fed into the game in a particular moment (for real-time games) or turn (for turn-based games). For instance, chess's input resolution is rather small: during each turn you can only move one piece from one position to another (whether that unit captures another unit is outside the scope of input; it's a part of feedback). Real-time games tend to have vastly larger input resolutions than turn-based games. In a single moment in *Quake*, you can start running, jump, turn 33 degrees to the left, go 12 degrees up, fire a weapon, and start shifting your weight midair in a different direction.

Modern games tend to go in the direction of the real-time model, with higher and higher input resolution, because it is erroneously believed that higher input resolution is better. This is completely untrue, and leads to many design problems. There are major pitfalls with games that have massive amounts of input resolution. First, they tend to be much harder to balance since the range of input possibilities is so massive that it's almost impossible for the designer to predict how powerful virtuosity could become in the hands of an extremely dedicated player. Second, these games tend to downplay strategic decision making and let execution take the lead. There are many examples of games—even games termed strategy games—wherein you can make the correct strategic decision to counter your opponent but lose because he or she simply passes input in faster than you. *Warcraft III* is a good example of this type of game, in which good players simply execute the same two or three strategies over and over regardless of what their opponents do. The game largely comes down to a match of "micro" execution.

To go back to the *Quake* example: most people automatically believe that the addition of being able to look up and down was strictly a good thing. Well, since FPS games had already started using different floor heights in levels, it definitely makes sense to allow players to look up and down. But I feel that few have considered that overall, gameplay may have been stronger if turning and aiming was only on one axis. How could this be? Because most of the interaction takes place on that axis anyway, and so the game allows a whole range of possibly unnecessary input information.

Video-game designers seem to forget that imposing restrictions on players is not a bad thing; *imposing restrictions is what game design is.*

We should be just as concerned about elegance when it comes to input resolution as we are when it comes to adding new features or mechanisms to our systems.

Feedback

Feedback is the term used for the opposite of input—*output*. It's how the game responds when you put information inside of it. In all games, this means feeding the input into the ruleset and outputting the result. In chess, if your input was "move to D4," the rules would require you to check to see if there was already another piece there. If there is and it's yours, the move is illegal. If there is and it's your opponent's piece, the piece is captured. Either outcome is feedback.

In video games, feedback is often a way to refer to the visual or audio representation of the actual feedback. The number 100 flying out of Mario's head is the *visual* feedback that shows players what the system fed back to them for their input: 100 points. It's important to know the difference between actual system feedback, and how a game visually or audibly represents it.

Execution versus Decisions

Some people believe that an execution barrier is a kind of decision, and that systems such as *Guitar Hero* are games because they include these barriers. I think that this is a very strange way to use the English language. If you're walking across a room and you trip on something and fall, and someone tells you that tripping was a "bad decision," you'd think that person was a jerk. Obviously, you never *decided* to trip and fall. In the same way, failing to hit a note in *Guitar Hero* is also not a decision.

To understand this better we have to go into the essence of the word *decision*. At the heart of the word's meaning is the idea that there is some unresolved question—that several options are in play, waiting for one to be chosen. If we're looking at a decision in the context of a game, we must first take for granted that the player is sincerely trying to achieve the goal (as per the rules of the game). If this is the case and there are two choices, one of which is clearly the optimal choice, then can you really say that there is a decision to be made? There is no unresolved question. It was already resolved before you began to even play.

It gets a little fuzzy because there may be ambiguity about whether or not you can execute something properly. In those cases, it may indeed be the case that some kind of execution barrier is providing variance (or randomness) to a game. There may be a safe route you know you can choose, and a harder route that you can't. This may be a valid decision

Haymaker Squat Punch (Ruby Level)

Figure 9. Here's how to do the Haymaker Squat Punch move—the game doesn't exist, but I'm sad to say it certainly could.

for awhile, but it will probably melt away soon as the player gains enough skill to at least attempt the optimal choice every time.

Many games that have a large element of execution are really contests. You can "choose" all you want in bowling, but if your opponent is capable of executing strikes 100% of the time, he or she is going to win. That is the clear optimal strategy in bowling. Execution has a habit of taking games over and stepping them closer to contests. If you're designing a game with an execution element, you have to be extremely careful with it and make sure that the optimal decision is always at least a little bit ambiguous.

Certain genres (I'm looking at you, fighting games!) have a very bad habit of incorporating crazy execution (and memorization) requirements into their gameplay. The thinking goes along the lines of, if a move is otherwise too powerful, balance it out by making it harder to input (see Figure 9). The problem with this thinking is that no matter *how* crazy the input is, eventually players will master it (see *1080 Snowboarding* or *Killer Instinct* for incredible examples of this) and your game will be thrown out of balance. Further, you're just making it harder for new players to learn it.

Randomness

Computers can't produce truly random numbers, and dice and cards cannot produce truly random results. However, their results are not predictable by humans, and so for us they're effectively random—and that's good enough.

Throughout this book I point out that a single-player game *must* have randomness in order to remain a game. Games without any randomness quickly break down and become memorization puzzles (*Castlevania, Super Mario Brothers*) or contests (*Guitar Hero, Dance Dance Revolution*). Without another human mind in play to try to throw you off, some kind of random information is required to preserve ambiguity.

Someone reading this book may get the idea that I'm very pro randomness in general, but this is not the case. In multiplayer games, I actually think randomness tends to be easily overused. At one end of the spectrum you have silly examples such as *Candy Land* (which actually is a contest—a luck contest—because there are no decisions whatsoever, let alone ambiguous ones), but even lower levels of randomness can cause problems.

One of the most popular Eurogames in America is *The Settlers of Catan* (see the "Designer Board Games" section for more about Eurogames). This is a relatively simple game in which you build a network of houses and roads, collect resources, trade, and perform a few other special actions in a race to get the most victory points. While the game is far from purely random, some people find the amount of randomness it does have to be a problem. You roll two dice on each turn, and whoever has houses on the corresponding numbered spaces gets resources. It's not terribly unlikely that you will go too many rounds in a row without your numbers coming up, and this feature can dramatically affect the course of the game: one player could have his or her numbers rolled ten times in a row, providing a massive advantage over the other players. These kinds of random ups and downs in games sometimes are referred to as *windfalls* (random good things) and *pitfalls* (random bad things).

The problem with a game that has a high level of randomness is that it can sever the tie between the player's *agency*—his or her decision-making abilities and performance—and the outcome of the game. It can send mixed signals to players who are trying to learn it (and for any unsolved game, that's *all* players), for they may do fantastically one time and poorly another, and the results may have much less to do with the choices they made than with randomness. Losses can feel like they weren't real losses, or that they don't matter, and wins can feel the same way. Games with a high level of randomness can have a bit of a lethargic "who cares?" feeling associated with them. BoardGameGeek user clearclaw remarked about *The Settlers of Catan*, "Just roll the dice for me, I'll go do something else and let me know how it turned out later."

Again, I'm not against randomness—but because of these kinds of problems you should never have more randomness than you need. Think of randomness as a necessary evil. The key thing to understand is that if you have strong mechanisms in a multiplayer game, you don't need very much randomness because the other players will provide most of the variance. And it's much more interesting and rewarding to fight another player than it is to fight a deck of cards.

Single-Player versus Multiplayer

Although it tends to be a useful term practically, there really isn't any such thing as a *single-player* game. Since all games are contests, all games must have more than one party that is trying to win.

But didn't I say that games are systems in which agents make ambiguous decisions? Are the computer-players really making such decisions? Well, actually, in some situations the answer is yes. Take a computer chess player, for instance: it doesn't know whether the move it's making is the optimal move, but it's making a kind of informed guess, just as a human being would. The computer is simulating a human opponent, so it's inaccurate to use the term *single-player* game. If chess is a single-player game, then any game in which a bot is the opponent is a single-player game.

However, not all single-player games work this way. Oftentimes, a computer-player is actually just rolling dice to produce its results. *Tetris* is a good example of this: here, the computer is certainly not making ambiguous decisions. So does that mean that *Tetris* is not a game? No! The opponent in *Tetris* is not the random-number generator. The RNG can never "win" in *Tetris*. The opponent is *you*—or whoever it is that currently holds the high score you're trying to beat. *Tetris*—and any other single-player score-based game[1]—is actually a multiplayer game that isn't played simultaneously. One player gets a high score, another player tries to beat it, and if he or she cannot, the match is lost. And so on. So in a way, score-based games are really multiplayer games that are played asynchronously, and because of this asynchronous nature, you can be your own opponent.

Survival, Completion, and Game Difficulty

First off, I should clarify that *difficulty* is not the same thing as usability or accessibility. All games should be as easy to learn and play as they possibly can be, but the reality is that some very complex or unusual games are almost certainly going to have some level of difficulty associated with learning to play. While we can minimize a lot of this difficulty with good design, it can't be completely avoided. Difficulty as it is discussed here is not the difficulty of playing the game but the difficulty of winning.

Let's start by getting a useful definition for *difficulty*. I think it would be fair to describe it as *the magnitude of the obstacle between the player and winning*. It doesn't make sense to apply this definition to multiplayer game systems, because for these games difficulty is based on how strong your opponent is. Therefore, *difficulty* as it applies to game systems (and as it's used here) is more relevant for single-player games.

It should be clear that without a goal, a system cannot be said to be difficult. In *Minecraft*, you can throw a million monsters at the player,

---

[1] Note that a game "having a score" and a game that is "score-based" are not always the same thing. Lots of games have a sort of vestigial score mechanism, while their actual goal is completion.

making it a horribly dangerous situation. But the system is still not difficult, because who said you were even *trying* to stay alive? Because it's difficult to *survive*, however, you might make the mistake of thinking that the system is indeed difficult. The thing is, who said that survival is the goal of *Minecraft*? Further, survival is not a valid win condition unless it's timed—otherwise, at what point can you say that you survived? An hour? Three hours? Twenty-four hours? Here, survival is an inherently unachievable goal in that it is logically incomplete. If there is literally no way to win, then there also can be no way to lose, because there was no contest to begin with.

Digital gamers everywhere have a tendency to assume that *survival* is the goal in games. It comes, in part, from designers' collective decision to make fantasy simulation the primary goal of video games. If you're really there—if you're actually putting yourself into the shoes of whatever it is you're controlling—then it sure would seem natural that survival is what you're setting out to do. And if it's really a fantasy simulator, then maybe survival *is* what you'll care about; but then again, maybe not—what if you're playing the fantasy of a suicidal person? Games need achievable goals to reach even the level of "contest" on the interactive spectrum, though. Survival can be that goal if a score is attached to it, but then *getting a high score* (which is achievable) becomes the goal instead of survival.

Another wrecking ball to the part of the brain that could otherwise understand difficulty is the assumption that *completion* is the goal of a game. Modern video games are unlike games created before the 20th century in that, like books and movies, the expectation for many is that they will be completed and then (for the most part) abandoned. Consequently, people tend to misunderstand the level of difficulty for games that are not about completion. For instance, Gamespot.com gave *Mystery Dungeon: Shiren the Wanderer* a 6/10 review (which is essentially a bad rating for a very good game, since most reviewers use the same scale as schools do: scores in the 70s mean *average*, and scores below 65 mean *fail*) mostly because the reviewer failed to understand the game. Editor Austin Shau said in the review that, "All told, *Mystery Dungeon: Shiren the Wanderer* is a largely frustrating experience because of its randomness and permanent deaths."

The key term to look at here is *permanent deaths*, which illustrates the reviewer's central failing in understanding this game. What *Shiren* actually has is just a *lose condition*, something that RPG fans and many other video gamers have sort of forgotten about. But in *Shiren* (or in any other games from its genre of *roguelikes*; see Chapter 4), death doesn't mean you lost. These are score-based games, and just like in *Tetris* or *Pac-Man* you win when you get a high score and you lose when you do not get a high score. Is *Tetris* also "a largely frustrating experience because of its randomness and permanent deaths"?

There was a failure to understand the goal of the game here, but to be fair, it wasn't entirely the reviewer's fault. *Shiren* didn't do a great job of telling players what its goals were—and it is painted up to look like a Japanese RPG, so I can almost forgive Mr. Shau for expecting to be able to grind and cruise his way through the game to completion. We cannot even begin to discuss difficulty without a solid understanding of an achievable goal. Because again—if a goal is not achievable, or if there is no goal, can a system really be called *difficult*? Once we understand what the goals of our games actually are, we can start talking about difficulty.

How Hard Should a Game Be?

In game design, *game difficulty* isn't actually a terribly useful concept. Allow me to explain why. Essentially, the process of game design can be broken down into two parts.

1. Adding rules to a system.
2. Balancing those rules.

If your rules are balanced, your game will have what I call a *balanced difficulty*. A balanced difficulty is a level of difficulty that (assuming equivalent skill and luck outcomes among all players) will provide each player with an equal chance to win. This applies to both multiplayer and single-player games. A balanced difficulty is an ideal, so there has never been, and will never be, a game that is truly, completely, and perfectly balanced. But it's important to get as close as you can! A balanced game is a great home for ambiguous decisions.

In the case of multiplayer games, of course, the biggest factor determining difficulty will be that of your opponent's skill. Some games, such as *Mario Kart 64*, like to tone down the effects of your opponent's skill level by using randomness. The more random a game is, the less player skill matters. The unfortunate thing about this approach is that the more randomness is added, the less players' decisions matter, damaging the feature that makes games special in the first place.

Bad Kinds of Difficult

Not all difficulty is created equal: some kinds of "difficulty" are based simply on winning a dice roll, or requiring the player to perform lots of uninteresting busywork to succeed. The *Fire Emblem* series is one example that

I like to point out of the bad kind of difficulty. These games are turn-based, tactical, top-down RPGs—normally, the kind of game that would interest me. However, there are too many little silly things required to actually succeed in the game. For instance, characters are easily killed (often after just two or three combats), and once they're dead, they're dead forever (unless you reload the game, which you will do often—another problem). There are random critical attacks, which often mean the difference between winning and losing a combat. Finally, the overmap doesn't always make it obvious what kinds of attacks a given character can carry out. "Oh, there's an axeman," you may observe. "I can hit him with my archer from a safe distance and not take a counterattack." But wait! Some axemen have a special ability to attack from a distance. You can find this out by scrolling over to the axeman and opening up his inventory to see what items he has. This means, of course, that you have to do that *every* time you see a new unit.

Combining a huge impact from randomness with a saving/loading system, as *Fire Emblem* does, makes no sense. The games are generally quite difficult, too, which means that a significant part of being good at a game is simply playing, rolling the dice, and reloading if you get a bad result. This is a bad kind of difficult.

Balance

What exactly does it mean to balance a game? It means that all possible actions are placed on the same levels of value to the player. A lot of people get thrown off by this statement because they think that I mean something like, all punches in *Street Fighter* should deal a damage of 10. What I actually mean is that if one punch deals 10 and another deals 50, there has to be something about the 50-damage punch that provides balance. Often this takes the form of a long cooldown (a period of time after you attack in which the animation is still playing and you're vulnerable if you miss) or a long warm-up (which allows players to see it coming).[2] Sometimes the balancing factor is cost. In *StarCraft*, for instance, a Battlecruiser is obviously a lot stronger overall than a Wraith. However, not only do Battlecruisers cost a lot more minerals and vespene gas to produce, but they're also farther up the tech tree. This means that there is both a resource *and* time cost to getting Battlecruisers that Wraiths don't have. It should be noted that players who spend all of their money on Wraiths late in the game will actually dominate players who spend all of their money on Battlecruisers, since Wraiths have excellent anti-

---

[2] Note that sometimes, moves are "balanced" by making them hard to actually execute. I think this is a mistake, as players will eventually get over that execution barrier and then the game will become unbalanced.

air-attack capabilities. This illustrates that balance is often irregular and conditional: a flow chart for illustrating the balance of units in *StarCraft* would be extraordinarily complex.

How to Balance Your Game

To balance your game, first (as always) begin with the minimal amount of content that can express the gameplay. Then, add content *only as needed.* Ignore what you think is "expected" for a video game; we have come to expect an amount of content that completely precludes balance. Be extremely cautious with asymmetrical forces, as they tend to increase the amount of balancing work required (read more about this in the "Symmetry" section).

You also should allocate time in the development schedule for balancing; this time will be used after the game is considered completely done. At a certain point you'll need to put new features on lockdown, and just play-test. Get friends to test the game. Find forums online and have people sign up to become beta testers for your game. Have them report what they find back to you, but don't make the mistake of trusting every balance report: many times people simply have bad games or confuse their own lack of skill for some kind of imbalance. Further, people generally will want you to increase the strength of their favorite things. Be wary of these kinds of recommendations.

As you find imbalances, it's generally good to have a light hand. Make the smallest changes you can to try to fix the imbalance, because large changes can cause all kinds of unforeseen problems. Also, know this: there is no shortcut to balancing a game. Fixes such as *dynamic difficulty adjustment,* wherein a game automatically adjusts its difficulty for the player, makes no sense and is a complete abomination of the purpose of games. This feature rewards bad play and punishes good play, which is obviously the exact opposite of what a game should be doing. Many companies seem to think that these kinds of Band-Aid approaches to game balance will work, but instead they make a game feel dead, lifeless, and undynamic. Other examples of such balance Band-Aids are *globalized leveling* in *The Elder Scrolls IV: Oblivion, rubberbanding* in *Mario Kart 64,* and the AI Director of *Left 4 Dead.*

Theme

As I mentioned in Chapter 1, a *theme* is a "literal" layer of information placed atop game mechanisms. Some games have very little in the way of a theme, and some games are chock full of it. What you'll be hearing from me again and again in this book is that you can't allow theme to dominate or damage your game designs—a problematic tendency of video game designers and Ameritrash board game designers (see the "Designer Board Games" section). You may ask: if themes cause such problems, why bother with them at all?

As with other elements, I would say one should include only as much of a theme as is needed. Generally, the more complex a game is, the more it may need thematic elements to help explain its verbs (actions). Themes make games easier to learn, although Go is fine with no theme, and probably couldn't be improved much by adding a thematic layer. On the other hand, *The Legend of Zelda* is inherently more complex than Go, and so there are elements that would be much harder to explain and remember without the theme. *Zelda's* bombs and arrows are an example of this: if the game was totally abstract, it would be very difficult to explain and remember the way that these items work. You'd have to say that *item X* (i.e., a bomb) can be placed on the spot where you are, and in three seconds it will destroy some other moving agents and possibly uncover new paths. *Item Y* (i.e., arrows), on the other hand, is shot straight from the angle you're facing and will destroy any moving agents they come into contact with. Without the thematic layer of information of bombs and arrows, players would find these rules arbitrary and very difficult to remember. So the theme helps players to learn to play the game (see Figure 10).

I have a personal story that illustrates something quite interesting regarding this subject. I've spent many years dabbling with emulators—programs that can emulate various video-game console technologies on other hardware. In this case, this was a Windows emulator for the Super Nintendo Entertainment System (SNES). A friend of mine and I were sitting down to play the Super Nintendo version of *Street Fighter Alpha 2*. Apparently this game used some funky technology that the emulator couldn't

Figure 10. This simple, iconic graphic of a bomb quickly relates a lot of useful mechanical information to the player. For this reason, it's helpful for a game designer to study symbology.

handle (or something) because when we loaded it up, the graphics were missing. The whole screen was solid colors, and the characters were represented by clusters of solid color hit boxes. Due to an error, *Street Fighter* had become visually abstract!

One might think that the game was unplayable in this state, but since we were both already seasoned *Street Fighter* players, we really didn't miss the theme at all. The game was just as playable, and might actually have been a little bit clearer (since hit boxes don't line up with character artwork 100%—hit boxes are always rectangular, for instance). The point is that once a player knows a game well you can subtract the theme, and as long as different things can be distinguished from each another the game will still work just as well.

Finally, I need to mention that the other reason to include a theme is attractiveness. This is a book about how to design better games, not how to increase sales, but a sad reality about the current world is that games (especially digital games) with a more abstract look can be a tough sell. Having a theme can be a great way to make your game feel more inviting, especially to more casual players. The challenge is figuring out how to add a theme without detracting from your gameplay.

Inherent versus Emergent Complexity

Understanding the difference between *inherent complexity* and *emergent complexity* is crucial for any game designer, and extremely useful for anyone who enjoys games. Because a game is simply a ruleset, the game (or perhaps the *match*, a session of a game) doesn't "happen" until someone plays it. Situations, problems, and patterns *emerge* naturally from the set of rules during gameplay. This complexity is therefore emergent complexity. This is in contrast to inherent complexity, which is complexity of the ruleset itself.

To illustrate, let's compare the two abstract strategy board games chess and Go. In chess, what are some of the basic rules? There are many, so let's look only at the rules for one piece, the pawn.

- Pawns can move one space, except when they're moving from the starting position, in which case they can move two spaces.
- Pawns can capture enemy pieces on a forward diagonal move.
- Pawns have a move called *en passant*, which allows them to capture enemy pawns that have just moved from the starting position by moving diagonally past them.
- Pawns can be promoted to other pieces if they reach the opponent's end of the board.

Now, these aren't even close to the complete rules of the game of chess—there are several other types of pieces and special relationships between pieces that make up the rules.

By most digital-game standards chess has a small number of rules. Now let's take a look at the complete rules for Go.

- The game is played on a 19 × 19 grid.
- The game is played by two players, one of whom plays the white stones and one of whom plays the black stones.
- Players can place one of their stones anywhere on the board, as long as (1) there isn't already a piece there, (2) their piece wouldn't be immediately captured, and (3) placing the piece won't restore the board to its previous state (known as the *ko* rule).
- If any number of your adjacent stones are surrounded by the enemy's stones, those stones must be removed from the board.
- At the end of the game, the player with the most territory and captures is the winner.

This is actually all of the information you need to play Go. There are a few other guidelines that you'd need to know to play professionally, but as you can see Go is a very simple game.

Or is it? While chess is certainly more complex *inherently*, Go is by far the more complex game in terms of *emergent* complexity. Mathematician Claude Shannon estimated that there are 10^{120} possible games of chess. While that's certainly a *ton* of games, the possibility space of Go dwarfs that at 2×10^{170}. Wikipedia says of Go:

> It has also been argued to be the most complex of all games, with most advocates referring to the difficulty in programming the game to be played by computers and the large number of variations of play. While the strongest computer chess software has defeated top players (Deep Blue beat the world champion Garry Kasparov in 1997), the best Go programs routinely lose to talented children and consistently reach only the 1-10 kyu range of ranking.

Kyu rank, by the way, is considered beginner to intermediate range, so all professional Go players can easily and consistently beat the greatest Go artificial intelligence. This is not just because of the possibility space, but due to the *meaningful* possibility space of Go.

Not All Possibilities Are Equal

Of course, real-time games played in real space, such as football or ice hockey, have infinite possibilities in a very literal sense. The question is, however, how many of these possibilities are meaningful inside the game space (endogenously meaningful possibilities)? In football, for instance, you can throw the ball out of bounds to stop the clock, a decision that quarterbacks sometimes make in a tight spot . You can choose to throw the ball forward and out of bounds, or directly to the side and out of bounds, or backwards and out of bounds. Technically these are three different possibilities, but they don't count as three distinct gameplay possibilities, because they all have the same *meaning* inside the game. Essentially, the entire out of bounds area is one discrete space in football, so it doesn't matter where the ball goes.

The reason people think Go has such a massive level of emergent complexity is not just because of the high number of possibilities. It's because of the very high number of *meaningful* possibilities. As game designers, we should be thinking about how to make as many meaningful situations as possible in our game systems. How one does this is dependent on the type of game, and creating these situations is really what's hard about game design. In future chapters, I'll give examples of how certain types of games can increase their possibility spaces in a meaningful way.

Hiding Behind Complexity

Some games overwhelm players with inherent complexity in order to keep them from seeing the basic dullness of a game. When there's 100 items, 40 characters, or 250 unit types in a game, many people just start thinking about inherent synergies and look past the core mechanics. It's plain to see that *Magic: The Gathering* would get boring fast without the vast amounts of inherent complexity in the form of thousands of collectible cards.

High levels of complexity are hard for players to see through, but it also means that they are hard for *designers* to see through. To make a game that really lasts, limit your inherent complexity levels so that you, the designer, can see any weaknesses in the core mechanism.

Information and Solvability

Perfect information and *complete information* are terms taken from game theory, which looks at logical decision making. Various elements of game theory are of varying levels of use to us—game designers. I recommend taking at least a casual interest in game theory, and the terms listed above are of particular importance to us.

We refer to a game as having perfect information when all players know all the rules of the game and everything about the current game-state. So chess has perfect information, because all players not only know the rules but can see everything about the current game-state. A strategy game without a fog of war system (which makes parts of the map outside of your unit's field of view impossible to see) has perfect information, but if the game has a fog of war, then it has imperfect information. Any FPS game has imperfect information due to the limited field of view (there are always areas of the game that you cannot see). Alternatively, when we say a game has complete information, this simply means that all players know the rules, but not everything about the current game-state.

In video games, designers rarely consider the question of information, and may change the type of information available to the player without realizing it. For instance, a game with scrolling almost certainly has imperfect information, because things are happening off-screen. Adding a mini-map with enough information may return it to perfect information status.

Solvability

It's wise to take some time and figure out which kind of information your game has, because it affects the *solvability* of your game. The first thing to consider is that *all games are theoretically solvable*—it's just a matter of how long it takes.

First, what does it mean to solve a game? It essentially means that we know the "answer" to the game, a strategy that is optimal. Sometimes this strategy is one that makes the player unbeatable, as happened with *Connect Four*. A mathematician by the name of James D. Allen solved that game in 1988, and his strategy allows the first player to force a win 100% of the time. Sometimes solving means forcing a draw (at the worst), as is the case for tic-tac-toe or Three Men's Morris. We all know this about tic-tac-toe; playing it with an adult almost surely leads to a draw every time. This is because we've all solved the game. However, for young children, tic-tac-toe can still be an unsolved game.

Games with imperfect information usually cannot be "solved" to the degree that the perfect information games can be. Since the game-state at a given point is not known, instead of an absolute solution, you get an optimal strategy. For instance, the theoretical optimal strategy of rock-paper-scissors is to play a completely random symbol each time. Of course, this may not actually be the optimal strategy in a limited context (i.e., in the way the game is normally played), but over a sufficient number of hands it would win. It may take a higher number of hands to start producing a better win percentage due to the natural proclivities of a human opponent.

Some games are partially solved. For instance, chess is not solved but there are various "solved" (i.e., optimal) starting moves (openings), and some solved endgames. Some versions of computer chess even take advantage of some of these endgame solutions if they come up. Go is much further from being solved, but even it has *joseki* (optimal safe starting moves).

The variable nature of the solved quality of tic-tac-toe speaks to a fascinating element about games: they depend on our mental limitations. It's entirely possible that some advanced alien species could come to Earth and find Go as simple to solve as tic-tac-toe. However, a game that would interest them would probably seem like sheer noise to us (in the same way that Go is not really playable for the child who can't solve tic-tac-toe). It's a balancing act for the adult person: the game designer has to work to create a system that is just out of reach of human mastery.

Symmetry

The idea of *asymmetrical forces* is very common in digital games, but interestingly, much less so in board games. The idea of an asymmetrical game is that you can start out the game with different forces (different available powers) than your opponents. Some examples of this are the different racers in *Super Mario Kart*, the fighters in *Street Fighter*, and the races in *StarCraft*. Before the match even begins, you're choosing your weapon.

Asymmetrical games are interesting in that they actually become many different variants of a given game system. Each matchup is really its own unique game, or variant. Those of us in video games tend to have very affectionate feelings for games with asymmetrical forces (I include myself, decidedly—in fact my first published game, *100 Rogues*, has asymmetrical forces). David Sirlin, an ex-pro *Street Fighter* player and board game designer, is a strong proponent of asymmetrical games and regularly talks about issues of balance and flavor (feel of use) with respect to asymmetrical games like *StarCraft*, *Street Fighter*, and his own *Yomi: Fighting Card Game*. Like me, he comes from a digital-game background, and so it's possible that he may be allowing his feelings for asymmetry to stop him from appreciating the inherent difficulties surrounding the concept.

Complexity

Physicist Albert Bartlett is quoted as having said, "The greatest shortcoming of the human race is our inability to understand the exponential function." And indeed, I feel that most asymmetrical game designers are failing to understand the exponential function with regard to their game designs. I'll use the popular example of *Street Fighter II* (I'm most familiar with the original SNES version of the game) to illustrate what I mean.

Let's imagine for a moment that we have a fictional version of *Street Fighter* that has only a single character, Ryu (a basic Karate Man type of character who has a super uppercut, a fireball, and a flying spin-kick move). The game consists entirely of Ryus fighting other Ryus, or *mirror matches* as they're known. First of all, realize that this absolutely would be a complete game—there would be nothing *wrong* with a Ryu-only *Street Fighter* game. If you think it would get boring, then there's a problem with *Street Fighter*'s overall design, because every matchup should be interesting, including mirror matches. Interestingly, many fans of asymmetrical games generally seem to think that it's OK if an asymmetrical game has uninteresting mirror matches.

So let's say the game, with its one character, has a rating of ten complexity points, a completely made-up unit of measure that loosely signifies how complicated the game-system is. We should assume that even under these circumstances, the game is interesting, and that therefore it will be quite difficult to design and balance (making interesting games is always hard, after all!). Let's assume that Ryu takes up eight or so of the ten complexity points, or about 80%. What does the remaining 20% go to? Well, there are health bars, a best-two-out-of-three match system, and a big rectangular stage with walls on either side. Of course, not all games put so much of their complexity into characters, but for *Street Fighter* I think that's the case.

What if we add in another character? How many complexity points would we be looking at now? That depends largely on how different the other character is from Ryu: if the character is just a recolor, the game is no more complex and the total number of complexity points would still be ten. If he's a character like Ken (who has all the same moves as Ryu, but slightly different stats), then we might have to take that character complexity (which was nine points) and add to it. If the character is different enough, though, we might be adding around four complexity points to our game. Adding two more characters, for a total of four, would make the game roughly twice as complicated as it was when we started! If your characters are significantly different from each other (which they should be—otherwise the reasons for introducing asymmetry are diminished, and the game simply becomes fuzzier and less clear), you can see that the amount of complexity in your game will skyrocket with 20 or more characters, like most fighting games have.

Balance (Again)

There are various mathematical equations that can help you figure out balance, depending on the various factors in your game. Continuing with the *Street Fighter* example, though, here's the way to do it. Let's say your game has ten characters in it. Using a math operation called a *combination* (you can find it by typing *choose function* into Google), we can try *3 choose 2*, which will tell us how many combinations there are (minus mirror matches). This number comes out to 3. For a two-player game like *Street Fighter*, we can simply add the number of characters in the mirror matches, bringing the total to 6. Note that you can't use that snazzy trick for a game that has to choose combinations larger than 2 (look up combinations on the Internet for more extensive information on these equations).

So for a game with three characters, we essentially have to balance six different games. Again, this is an interesting property that we have to remember about an asymmetrical game—it's essentially several games you're developing at once. When players agree to play the Ryu vs. E. Honda game, it's a different game from the Ryu vs. Ken or the Ryu vs. Ryu game. Each combination is going to require its own balancing job. For a game with ten characters, you have to balance *55 games*. How about a game with thirty characters (which, I might add, is kind of the standard these days)? That game actually consists of 465 different games that have to get balanced! That's insane—do you think you'll ever make 465 completed games in your entire lifetime? And we wonder why fighting games are always imbalanced, with tier lists (rankings) of characters that lay out which are the "good" (most powerful) characters regularly being created.

Now some of those 465 games may be easier than others, if they are very similar to each other. However, making a game with nothing but very similar asymmetrical forces is almost always a very bad idea: a lose-lose situation. You're making your job harder, and providing very little of what people like about asymmetrical forces. Further, you're blurring the choices (making them closer to false choices) by making them similar.

In the end, the solution may be as simple as this: asymmetry can be fantastic, but keep the number of races and characters low. And video gamers: by low, I don't mean a dozen—I mean preferably less than half a dozen. This is one of the ways that the computer has spoiled us digital gamers—we have to really reset our expectations now in a dramatic way if we're ever going to make something that stands the test of time.

False Choices and Other Sins

Games are all about choices. Their interesting, ambiguous decisions and the consequences that result are a fundamental part of what makes games *games*. So what does it mean for a game to have *false choices*?

Video gamers are all too familiar with the false choice, which is blended in seamlessly with the larger *non-choice* experience. *Half Life 2*'s level design was, in fact, lauded for how convincingly it went about the matter of presenting false choices. People said, "even though it's totally linear, it never really feels linear." When a building collapses over a door, leaving only another corridor as an available route, you feel as if there *was* a choice to be made, but you just barely missed it.

Any choice in the category of "do X or die" is also a false choice. Of course, almost all games have some of these choices. In chess, for example, moving your king out of checkmate is a false choice (sometimes you have a choice about *where* to move him, but not always). In a checkmate situation, the rules literally forbid you from doing anything but moving your king. In every game there will be similar situations.

Saving/Loading

Many false choices are easily avoidable, and we should take care to do so. In my opinion the typical savegame system, which lets you save the game at specific points (or worse, at any time) and reload it at any time, turns *everything* in games into false choices. If you don't like the outcome of a given choice, you can (and really should) simply undo it by loading the game. For this reason, saving/loading is a form of legalized cheating, and something that needs to be addressed very soon in mainstream digital games.

All of the reasons for the defense of saving/loading come from game features that are problematic to begin with. "We need saving/loading, because the game has to be 80 hours long!" "We need saving/loading, because the game is story-driven and therefore the player has to win!" "We need saving/loading, because we couldn't possibly balance this amount of content!" This is the nature of the defense of this mechanism.

Keep in mind, I'm not against a system that lets a player suspend the game—I'm not saying players should have to play games in one sitting. Smart games, such as *Mount & Blade* and most roguelikes, have save-and-exit systems, allowing players to stop playing whenever they want and then continue from where they left off. Loading is only possible from the title screen. I'm advocating that *all* games have this system.

Grinding

Grinding (as defined by ex-*Dungeon Crawl: Stone Soup* lead designer David Ploog) is any activity that a player can do repeatedly with minimal or no risk and that results in an in-game reward. The most obvious and well-known examples of grinding are those found in RPGs (particularly Japanese RPGs), in which you can (and often *have to*) fight low-level monsters (who are no threat to you) over and over and over again to gain experience points. But there are other examples in other genres, too: one such example might be fetch quests, which have you run an item from one place to another in-game. Almost all of the gameplay in *Farmville* is pure grinding. The game tells you, "click on this thing for 300 points." There's no reason to *not* click on it, so it's just a chore the game makes you do.

Grinding is bad for two reasons. First, it presents a false choice because you *should* grind your *Pokémon* up to level 99. There's no in-game reason for you not to do that: it only benefits you. This leads to the second reason, which is that you're *motivating players to bore themselves*. In fact, you're using the player's boredom to counterbalance the tedium of grinding. The thinking goes, "Well, players will become totally unbalanced after grinding for two hours, but they'll get bored before that."

Perhaps this begs the question, why do so many games have grinding if it's obviously such a bad thing? The answer is that grinding *works*. The human mind is an archaic and exploitable thing, and our evolutionary imperatives are easily taken advantage of. Evolutionary needs, such as the need to gather and the need to show status, are being exploited when we're playing *Pokémon*, *World of Warcraft*, or slot machines. No one finds grinding interesting; it is a compulsive behavior.

Game designs should never encourage people to do uninteresting no-brainers once, let alone repeatedly, because games are about decisions and building skills. Games can, and should, be making every effort to enrich the lives of their players, and not simply suck their time away from them.

Too Many Choices

The opposite of the false-choice problem (which is not having enough choices) is to present the player with too many choices. Having too many choices can be just as bad as having too few, and like so many other things in life, this element of game design is a balancing act.

A very obvious example of a game with too many choices is when your hand consists of 20 or more cards in *Magic: The Gathering*, but in most cases it's not obvious when a game gives players too many choices. The example from *Magic: The Gathering* seems obvious—it just feels like too many choices for the player. But why? What's the harm in too many choices?

In turn-based games, the most obvious and apparent problem with having too many choices is something called *analysis paralysis*. This condition manifests itself clearly in multiplayer board games, as other players can wait a long time for the current player to take his or her turn. In other games, it's a lot less clear when there are too many options. The damage done by having too many options is usually that each option starts to lose uniqueness, and the game starts to blur into a somewhat arbitrary guessing match of tiny decisions that have little meaning. Much of the *weight* of decision making is lost under these conditions, even though it's less apparent.

This relates to our earlier discussion about the essence of design: elegance. Elegance is doing a lot with a little, and giving players dozens of options for each thing they can do usually entails a lot of inherent complexity. Having 20 guns, or cards, or moves that a player can do is probably a bad thing. For instance, in *Street Fighter* each character has dozens and dozens of moves, all of which can be performed at any moment during play. Extreme *Street Fighter* fans will tell you that every single one of those moves matters—and at the highest levels of competitive play that's probably true, as any small advantage can mean the difference between winning and losing—but the reality is that many of the moves serve the same purpose to varying degrees. The result is that many of them never get used, because they do the same thing less well than some other move.

This speaks to the real problem with having too many options in a game: in any given game, there are only a small number of meaningful things you can do. In *Street Fighter*, for example, your real choices might be attacking high, attacking low, blocking, throwing, jumping, and shooting a projectile. But *attacking high* alone includes ten or so different options for performing it. The reality is that at least a few of those options will be simply not as good as some of the others in performing the *real* choice of *attacking high*.

Good game designers understand what the real choices in their games are, and usually limit the number of in-game choices to be similar to the number of real choices in the gameplay. They know that when a game has too many choices, many of them will be false choices.

Efficiency

Do not, under any circumstances, waste *any* of your players' time. Players' time should be absolutely paramount, and you need to be doing everything in your power to deliver as many interesting, cool decisions to them as you can per second of play. Keep in mind that they're probably quite busy, and have taken not only some time out of their days but also some money out of their wallets in order to check out some of your ideas. Giving them chores or waiting times is completely unacceptable. If you have screens that take a long time to load—and by long, I mean more than a second or so—find a way around it! It's so important that players not wait that art and programming should be restructured to avoid long loading screens.

In video games, players' time is most often wasted via no-brainer actions such as grinding or other false decisions. Level geometry is almost always much bigger than it needs to be, forcing the player to run down long hallways over and over again. If your game includes something like a town with various NPCs (non-player characters) whom you can talk to, ask yourself what the purpose of the town is. If its only purpose is to buy and sell items, for instance, perhaps a text menu could replace it. Let the player get right to the meaningful stuff.

Board games are usually a bit more respectful of players' time, due to physical constraints and the fact that adding wasted no-brainer actions makes games annoyingly fussy, but there are still examples of places where they can tighten up. The genre most frequently responsible for being inefficient is the war game, which tends to have a good amount of no-brainer actions and maps that are too large. This is done for the sake of simulation, but if you look at these games through our lens, they come up short in this area.

Take Nothing for Granted

FPS games have guns. Platformers have scrolling. Fighting games have asymmetrical forces. Dungeon crawlers have loot. Video games have achievements, cutscenes, quick-time events, RPG elements, chest-high walls, combos, etc. People on the Internet debate the characteristics that make a good boss fight, as though all games are so similar that what works for one game should work for all (which, sadly, is almost true right now). One of the assumptions that I dislike the most, and that I think many indie game developers are guilty of, is that video games should have jumping. But it doesn't have to be this way: you should start from scratch and ask difficult questions no matter what kind of game you're making. Creativity sometimes means being destructive—destroying old ideas and expectations, and building new ones.

I think that my game *Auro* is a good example of this kind of creative destruction. It's a turn-based, randomly-generated dungeon crawler. When I started designing it, I wanted to make a roguelike game that was similar to my previous game, *100 Rogues*. But soon after I began the design process, I started to see really massive, fundamental ways that I could change things.

The first thing to go was equipment. Roguelikes and RPGs always have a system for you to find or buy equipment for the characters. The problem is that this system is inherently unnecessary: it's just an extension of your level (at level 10 you get the level 10 sword, and then you have to equip the character; at level 20 you get the level 20 sword, etc.). Of course, if there aren't level requirements and a level 1 character can equip a character with a level 20 sword right off the bat, then forget about balance—it's just a big mess. Instead, I decided to render equipment like swords and armor as special abilities you could get in a special discipline skill tree. So you just take the *sword* skill, and bammo—you have a sword!

Then I started thinking about the classic experience and leveling-up systems. Most RPGs have a huge problem in that infinite grinding is possible (and not only possible but required in many of the games). Roguelikes put a cap on grinding through the food system, which, while functional, has never really satisfied me. This system works by having a food clock that is constantly counting down, and you either have to eat or starve to death. Usually, food can only be found by moving forward in the dungeon. There are some problems with this system: it's relies too much on randomness to give the player food, and it only puts a soft cap on grinding instead of preventing it.

I also started to question the idea that you get stronger as you go. In most of these games (my own *100 Rogues* included) your characters' stats grow as the game progresses. But does this really make sense in terms of game design? The game should be getting more difficult, so why am I making my job a million times harder by having the player's power change over the course of the game? If anything, in some ways all RPGs tend to get easier as they progress because of the increasing stats. I'm sure we can all think back to many RPG experiences that ended with our characters simply being immortal demigods: this is the logical conclusion of a system that makes your character get better as you go.

So I ditched that, too. Your characters stay at the same levels of health and attack damage as they go through the game. Instead, they just learn new abilities as they go and face more monsters with an increasingly wide array of abilities. I also ditched the scales that monitor health and damage throughout the game—you can do that if characters aren't simply getting more powerful. Once I did that, it made sense to ditch those sword and armor skills as well and swap them out for more interesting abilities that expressed something similar in a way that was deeper and more sensitive to the context. For instance, I added an ability called *counter*: you cast it, and the next time you're hit with an attack or special ability (or both) they're combined and reflected back on the opponent. This is an inherently more interesting way to render *reduction of incoming damage*, as opposed to the usual platemail armor that just reduces incoming damage across the board.

Ditching increasing stats is a great example of taking nothing for granted. I think that when you take this approach to game design you'll find many things about modern video games that simply do not make sense. The classic idea of leveling up was included for purely thematic and fantasy simulation reasons, but we're making games here.

Becoming an Expert

No matter what kind of game designer you are, you can learn from other kinds of games. A digital-game designer has a lot to learn from board games, sports, and even other disciplines that aren't directly related to games. One of the most damaging effects of the isolationism of the digital-game industry is duplication. Most of us in the digital-game world tend to think of digital games as the only *real* games, and this explains why so many modern digital games are so similar to each other. We can't imagine something that we've never seen before. Creativity is the act of combining and modifying things already seen or experienced—you cannot create in a vacuum, you can only create using the pieces of information that you have available to you. The rewards of becoming an expert are huge for a game designer.

If you want to be a game designer of any kind, you'd be well advised to be an expert not just in digital games but in all kinds of games. In fact, I'd go so far as to say that you are drastically limiting yourself as a game designer if you aren't looking outside of your immediate medium.

The following are some examples of areas that all game designers should be looking at. I also recommend reviewing the games mentioned in Chapter 3.

Designer Board Games

This is the single greatest category of games available for someone who is interested in learning about new game mechanisms. Designer board games are a celebration of the concept of *game* as I define it. Since the 1990s there has been something of a renaissance in the world of designer board games, which may be related to the fact that their creators put a lot of pride into the practice of game design. Indeed, these games are known as *designer* board games because of their practice of including the game designer's name on the front of the box. This category includes many subgenres, only some of which I will address. In Chapter 5 I'll go into much more detail about these genres.

- ◆ *War games.* This is one of the oldest genres of games, as I mentioned in Chapter 1. War games tend to be extremely complex, long, and gritty. They usually straddle the line between wanting to be a game and wanting to simulate a conflict, and commonly are set in a real place or even simulate a historical battle. The nature of war games tends to keep them from being among the most elegant of games, but there's certainly a lot you can learn from them. The war game genre also includes the subgenre of tabletop war games, such as *Warhammer 40K*.
  - Check out: *Advanced Squad Leader, 2 De Mayo*, and *A Few Acres of Snow*.

- ◆ *Abstracts.* Abstract games are games that have minimal or no theme. They're entirely representational, and usually use a grid and basic shapes in basic colors to visually represent territory or other mechanisms of a game. They're often two-player games and their play frequently has low levels of randomness. Abstracts are especially useful for game designers because of their complete and total focus on mechanisms. If an abstract game's gameplay isn't at least slightly new or interesting, it sticks out like a sore thumb. That said, abstracts can be a bit more difficult to get into than some other kinds of games due to their lack of themes.
  - Check out: Go, chess, *Arimaa, Blokus*, and *Hive*.

- ◆ *Eurogames.* This is my favorite category of board games because they are heavily mechanical like abstracts, but with usually just enough of a theme to draw players in. Most of the notable Euros tend to come out of Germany, which is generally considered to be the board-game capital of the world. These games often have themes relating to farming, trading in the Mediterranean, or the medieval period, but what's notable is that Eurogames tend to specifically avoid direct player conflict, putting them in stark contrast to the category of war games. Another thing I personally love about Eurogames is that they are very elegant and often minimize the element of luck (few Euros have dice). In terms of elegance they perhaps are one step down from the abstracts, but they also tend to do a lot more than most abstracts.
  - Check out: *Puerto Rico, Through the Desert, Agricola*, and *Caylus*.

- *Ameritrash.* The term *Ameritrash* started out as a pejorative term, but eventually was embraced by the community and has now mostly stuck. The name came about because the games being described were mainly of American origin with a lot of components. Ameritrash games tend to ship with a whole lot of plastic. They also tend to be thoroughly thematic—even driven by theme in many cases. Almost every Ameritrash game comes with at least a few dice, and many come with a dozen different dice to use in different situations. Ameritrash games seem to have evolved out of the tabletop and pen-and-paper gaming fields, oftentimes being more tightly packaged *D&D* (*Dungeons & Dragons*) or *tabletop lite* experiences, but they've also come into their own in recent years. It's worth noting that Ameritrash games seem to have the most in common with modern video games: heavily thematic experiences with a big focus on production values.
  - Check out: *Arkham Horror, Battlestar Galactica*, and *Chaos in the Old World*.

Card Games

While there is some overlap between this category and that of designer board games, there's also a family of games that are played with regular playing cards. In fact, this category of games is so huge that going into it is well beyond the scope of this book. Subgenres of card games include trick-taking games, bidding games, and gambling games, to name a few. It's a genre of games with an extremely rich and vibrant history. I highly recommend Scott McNeely's *Ultimate Book of Card Games: The Comprehensive Guide to More than 350 Games* as a wonderful starting place for those who want to get more acquainted with the world of card games.

- Check out: *Tichu*, euchre, poker, and Reiner Knizia's *Money*.

Pen and Paper Games

There is a huge world of pen and paper games, quite possibly because they are easy to make (at least in terms of the material needs of production). I recently visited The Compleat Strategist, a game store in New York City, and was completely stunned by its collection of books, manuals, maps, and other materials relating to pen and paper games.

Pen and paper games often attempt to do something that I don't think is the job of games: to simulate a world or interpersonal experience. With that in mind, game designers can still learn something from P&P games since they are interactive systems that usually do have goals and are often very game-like. They're a great place to look for inspiration for a new digital-game mechanism. As I pointed out in Chapter 1, they're the bridge between board games and digital games—you should know about these games for that reason alone.

- Check out: *Dungeons & Dragons*, *Pathfinder*, and *Paranoia*.

Sports

I don't need to explain the social significance of sports in the life of a modern human being. You can't go anywhere without seeing professional baseball games on television or seeing something about a football victory in the newspaper. And yet, many of us draw a somewhat arbitrary, clear line between games and sports. Those of us in the game design world tend to think sports are not really games—or if we do consider them games, we don't really think to analyze them. But the fact is, we all have a lot to learn from sports.

Sports are probably the oldest form of game playing, and depending on how you define *sport* (we all agree golf is a sport, right? Then is croquet a sport? How about billiards? Skee ball?), the range of mechanisms is huge. Moreover, people don't realize that there are officials who address balance issues and rule changes for sports every year, particularly in the case of American football.

A look into the history of how the rules of American football evolved over the years yields an enormous number of lessons for game designers. For instance, Rogers Redding, who officiated NCAA football for many years, has talked about some of the unintended consequences of one particular rule change in that game: requiring hard shell helmets. He believes that while this rule was added with the intention of protecting players, it actually may have had more effects than that. One effect may have been that players now move faster and more recklessly. The irony here is that because players feel well protected, they may make worse decisions at higher speeds, causing more serious injuries. As game designers, we all know too well that adding one new rule can have the same kind of unforeseen consequences.

Game designers can also learn a lot from the structures and measuring systems used in professional sports, such as the different types of tournament and league setups, the ranking systems, and the systems for metrics. Learn to play a few sports—and don't rule out sport-based video games, one of the last bastions of true gaming in digital games today!

- Check out: football, soccer, tennis, and golf.

Children's Games

While most children's games will not fascinate or capture the imagination of the average adult, it's definitely worthwhile to know your stuff about them. One reason to know about children's games is that you can see very clearly how games can break down depending on the intelligence of a user. We all had fun playing *Candy Land* as young children, but it's hard to see where that fun came from now that we have minds that are more developed and a better understanding of game mechanisms.

Another reason to know about children's games is to absorb their lessons of simplicity and ease of use. Children's games don't do a lot, but what they do, they do with elegance, since this is of utmost importance with children.

+ Check out: *Chutes and Ladders*, *Candy Land*, and *Whoowasit?*.

Video Games

I think a lot of us think of ourselves as video-game experts, but there is really so much to know about video games beyond what is mainstream. Look to old DOS, Amiga, Apple II, or Commodore 64 games. Look up weird Korean web games. Visit abandonware websites like Home of the Underdogs and click on games randomly for hours on end. Read the articles posted on great game-history sites like Hardcore Gaming 101. The Japanese have had a habit of not releasing some of their best games in North America for the last 20 years, and looking at these games can also be extremely eye-opening. There are even some very worthwhile television shows and YouTube channels that should be looked into. Less well-known but rich genres such as MUDs (multi-user dungeons) and roguelikes are other examples of games that have a lot to teach us. And don't forget to check out Chapter 4, where I go into detail about various video-game genres. There's a ton to learn about video games beyond popular modern console and PC games.

It should be obvious, though, that no one ever fully becomes an expert. Nevertheless, it's an ideal that we should work towards: the important thing is to make a conscious decision to go down that path.

Related Disciplines

If you want to become a great game designer, it's worth taking at least a brief look at related disciplines. Some of these fields are more related to game design than others; the ones listed next are from most related to least related.

Psychology

I'm far from an expert on psychology, but I know enough to know that it has a strong relationship to the field of game design. Psychology is the study of human behavior, and games are machines for human behavior to attach to, so it makes sense that learning about psychology will benefit the game designer.

One area of psychology that's helpful for understanding games is compulsive behavior. Many popular video games, such as *World of Warcraft*, *Pokémon*, and *Farmville*, exploit human psychology to keep people playing. These games take advantage of our deep desire to gather (*Pokémon*), show status (*World of Warcraft*), and receive rewards (*Farmville*).

Many have used the term *Skinner box* to refer to these types of games. This is another name for behaviorist B. F. Skinner's *operant conditioning chamber*, which was literally a box with a button inside. Pigeons or rats would be placed in the box and receive food when they pressed the button (in some studies an electrified grid was activated). This created a clear pattern of consequences, which in turn modified behavior (a process behaviorists call *operant conditioning*). Many of you probably already see the obvious similarities between this box and, say, *Farmville*, which is all about constantly rewarding the player.

One could counter that all games manipulate human psychology. While this is absolutely true, the question is, what are players getting out of a game? Good games have all kinds of great, practical benefits for the human mind that go beyond pure enjoyment. Great games can be enriching in the same way that fine films, albums, or paintings can be, but so-called skinner box games tend to leave people empty, since they are *merely* being exploited.

Game Theory

Game theory is not the same as game design theory, which is the subject of this book. Game theory is a very formalized logical science that attempts to predict behavior given certain game-like situations. There are some great online resources for learning about game theory on the Internet, and it's definitely worth learning a thing or two about it. Some of the "games" of game theory may even inspire an interesting game, but more likely, you'll find the process of working out these dilemmas parallels that of working out game balance.

One of the simplest and most famous examples of game theory is the *prisoner's dilemma*. It goes like this: two men have been arrested and taken into separate rooms to be questioned, and each prisoner has the choice of whether to betray the other. If both prisoners choose not to rat

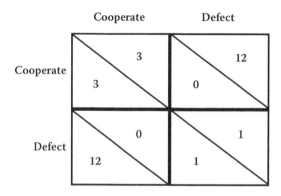

Figure 11. A chart of the prisoner's dilemma; the numbers stand for the number of months in jail associated with each alternative. (The numbers used are arbitrary as long as the relative values of each outcome are maintained.)

the other out, they each get one month in jail. If they both betray each other, they each get three months in jail. If one keeps quiet and one rats the other out, the one that snitched goes free and the other one gets an entire year in jail. Dilemmas such as these are often laid out in charts like the one in Figure 11.

Using a chart like this to lay out all possible outcomes and assign a number to the value of each outcome for a given player can be an effective process for balancing games, since this method allows you to quantify values for different moves in your system. Let's quickly draw up a chart (Figure 12) that could be applicable to any competitive fighting game—perhaps we could call this one the *Street Fighter's dilemma*.

This is, of course, a very simple example, but the chart does allow you to see how the two choices—attack or dodge—seem to work. The

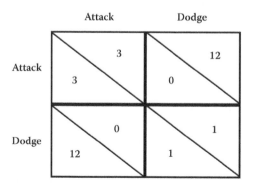

Figure 12. The *Street Fighter's* dilemma.

player that dodges the other player's attack can counterattack and gains the most points. A scenario in which both of them attack yields a little less value, and both of them dodging each other yields almost no value. It's not one-to-one, but you get the point.

If you study game theory you'll find deeper, more interesting dilemmas that are tougher to solve. And the thinking associated with solving them is much like the thinking a player must do to excel in a game—that is, predicting opponent behaviors.

Symbology and Graphic Design

If your game includes graphics—and most (but not all) video games and board games do—then you need to study graphic design and symbology. You'll be surprised not only at how difficult it can be to express certain game actions with an icon, but also at how much explanatory power a good icon can have.

Graphic design can help teach you how to organize your information on your HUD (heads-up display) or game board. A game that is well designed graphically is simply easier to play. You could always hire a graphic designer to help you with this, but the advantage of learning some of it yourself is that you'll sometimes find that a system's graphic-design problems can be a sign of game-design problems. If your information is really hard to organize, there may either be too much of it or it may not be related closely enough.

Teaching Your Game

The greatest game in the world is of no value to anyone if people can't figure out how to play it. In ancient times, games were taught via oral tradition: one generation passed down the rules to the next, and along the way (as tends to happen with such systems) there was probably some random mutations. The invention of writing allowed for the invention of the game manual, which allowed for more complex rulesets. More recently, the interactive tutorial has been introduced as a way to show people how to play video games. Regardless, teaching your game is going to be difficult, and the more interesting and new your game is, the more difficult it will be. Manuals have the advantage of being completely voluntary for the user, and can be very in-depth, but they can have the disadvantage of being harder to understand and requiring more focus. Interactive tutorials allow the player to play while learning, which can be good, but many games make the terrible mistake of forcing tutorials on players by making the first hour of play a tutorial that can't be skipped.

I recommend that all games do what they can to teach the player about how they work *naturally* through *invisible tutorials*. You can give players tons of cues to help them figure out how to play by using smart graphic design, level design, character designs, or even just well-placed and clearly worded in-game text. A great example of the invisible tutorial is the first level of *Super Mario Brothers*, which blogger Anna Anthropy outlined in great detail in her blog post, "To the Right, Hold On Tight."[3] Not all games are as simple as *Super Mario Brothers*, and so most games can't completely teach their mechanics on the fly. This is OK, of course! But we should do what we can to teach what we can about our games in a hands-on way.

Conclusion

For more specific suggestions and ideas about game design, I recommend reading Chapters 4 and 5, wherein I analyze various genres of video games and board games and make suggestions about how they could be improved. Entire new genres can be created by asking difficult questions and bucking conventions that don't make sense. But above all, it's most important to always be designing. I've recently taken to keeping a pen and paper by my bedside, and every night before I go to sleep I quickly design a game. Often I'll just start with a simple title—sometimes simple ones like *Dragon Duel* or *Monster Basher*, and often absurd ones like *Pan Butterer* or *Penguin Lords*. It's a good challenge to start with these vague ideas and see what kinds of mechanisms you can dream up for them. And you never know—*Pan Butterer* might become a fantastic finished game someday!

---

[3] See http://www.auntiepixelante.com/?p=465.

3

How We Got Here

I don't think I need to go on very long about the value of an intimate understanding of the history of games—the word *history* itself comes from the Greek word meaning *knowledge from inquiry*. This is not a history book, and this is not a history chapter, so this chapter is limited to a brief review of a few famous games, styles, and movements. We'll develop our lens by using it to make determinations about various elements in game history. For a more in-depth look at specific games throughout history, I highly recommend R. C. Bell's *Board and Table Games from Many Civilizations*.

Ancient Board Games

It amazes me to look at games from thousands of years ago and see that even in those primitive times, there existed game designers who had an understanding of the properties important to a good game. In some games, you can see that these designers were already building abstract systems that had one intended purpose: allowing players to make interesting decisions. All of the games, contests, puzzles, and other interactive systems covered in this chapter have at least some interesting elements to them, and you would do well to find online versions of them, or download print-and-play versions of as many as possible.

R. C. Bell divides ancient games into six categories: race games, war games, positional games, mancala games (also known as sowing or count-and-capture games), dice games, and domino games. For our purposes, his categories will suffice as a starting point for understanding some common types of games.

Race Games

These are usually systems highly influenced by luck that usually feature some kind of circular track that players must race around to win. The tracks most frequently have a circle-and-cross formation, allowing for shortcuts.

Two major features of ancient games are reinvention and evolution. Almost every game has another, earlier game that laid the groundwork for it. This is so clearly the case that if we were to find an utterly unique game somewhere, it would be reasonable to say that we simply have not yet found its predecessor.

One popular example of a race game—and a good case study for evolution—is the modern-day *Sorry!*. This game is actually a modern version of an 1896 game called *Ludo*, which is based heavily on the much older Indian game of pachisi, which has itself been reinvented as *Parcheesi* (Figure 13). Pachisi (also called Chaupar) dates back to about 400 AD. However, it really doesn't end there; there are many earlier race games that clearly influenced pachisi.

Pachisi   *Ludo*

Figure 13. Early race games.

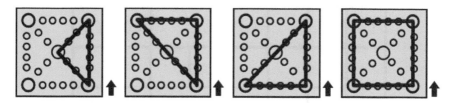

Figure 14. Possible nyout routes.

Some of the earliest race games were the aforementioned circle-and-cross games. One of these is a Korean game called nyout (or yut nori), and it could date as far back as 50 BCE.[1] Nyout is played on a square or circular board with two crosses going through the middle (Figure 14). Players roll dice (actually, they didn't have dice at that time and instead used marked sticks that would be thrown to get random results) to move along the track, and the first person to return to the starting tile wins the game. Players who land on the center tile or one of the corner tiles are allowed to take the shortcut.

Most people today would find nyout and related games to not have nearly enough decision making, since they are literally driven completely by luck. The modern *Sorry!* has at least one decision—choosing which piece to move—but earlier circle-and-cross games had no decisions to make at all. For this reason I consider games like nyout to be contests, not games.

In the race category, we also have many backgammon-style games. One of the most famous—and also one of the oldest games on record—is the Royal Game of Ur, also known as the Game of Twenty Squares (Figure 15). Like nyout, this is a game where you simply roll dice to move your pieces and try to get to the end of a track. Also like nyout, the player has almost no choices to make during play. Many believe that this game (and possibly many early games) was played for money, and there's evidence that not only was the Royal Game of Ur played with an initial starting wager, but that landing on one of the four corner points (labeled with hexagons) forced players to increase their wagers. So while the game itself does not involve decision making, there is the meta decision of how much to bet. This feature would be developed and result in a new category—gambling games—which we'll address in a moment.

It should be clear how both of these games evolved to create the still tremendously popular backgammon, which also has a mechanism based

---

[1] Interestingly, nyout has been useful to anthropologists: although it was created in Korea, it was later found featured in carvings and other representations in Mayan ruins, suggesting that North America was indeed populated from northeast Asia.

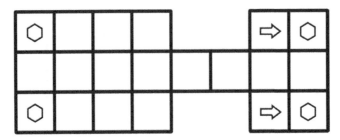

Figure 15. The Royal Game of Ur.

on dice rolling and racing around a track. A Roman game called tabula bridges the evolutionary gap in a very obvious way.

I started with these ancient race games not only because they are some of the oldest games we have on record, but because they show a very distinct path of evolution. The question is, what are they evolving towards? We'll revisit this question at the end of the chapter.

Gambling Systems

Most of the games listed so far do not hold up tremendously well for the modern adult mind. In fact, some of these games provide literally no choices for players. As I mentioned in Chapter 1, this renders these interactive systems *contests*: they simply measure luck, for lack of a better term. Modern games such as *Candy Land* and *Chutes and Ladders* are akin to these games, and adults rarely use such systems. However, adults do engage with such systems when there is money involved, which was the case with many of the games listed above. From slots to roulette to the lottery, there are many popular games that give players few or no meaningful choices, yet adults still play them actively and with great excitement.

Since gambling is not only a popular pastime but also a hot topic both legally and morally, there has been much research on the psychology of gamblers. A big reason that gambling games make any sense at all is because of something that psychologist Ellen Langer calls *the illusion of control*. We see examples of this in how players carefully and thoughtfully select lottery numbers, breathe on dice before rolling them, and engage in other superstitious behavior. Of course, none of these things actually have any effect on the outcome of such systems. So why do we do it?

One reason may be that an evolutionary survival mechanism is at work in a place that it shouldn't be. In the course of human evolution, it was to an individual's advantage to see agency behind ordinary events.

For instance, the rustling of a bush may very well not have been caused by another (competing) human or a dangerous predator—but those who learned to see it that way ended up with a higher chance of survival. It may be that much of our superstitious behavior and even many of our beliefs come from this survival mechanism. Regardless, the important point to take away is that gambling systems such as nyout, in which the player has no meaningful decisions to make, are contests, not games. This is not to say that all gambling games are contests, however. Games such as poker are certainly games, and of course any game can be turned into a gambling game simply by wagering money on the outcome.

War and Territorial Games

Alongside these luck-based gambling systems, you also see the emergence of war-themed games. Some are more abstract than others, and all are usually grid-based strategy games. These games, along with traditional sports, form the foundation of our modern understanding of games and my philosophy about what a game is.

One cannot write a book on game design without mentioning the classic game Go, also known as *weiqi* or *baduk*. This is a territorial game in which players place black and white stones in such a way as to capture areas or groups of enemy stones. What's utterly fascinating about this game is what an incredible amount of strategy goes into play. The game

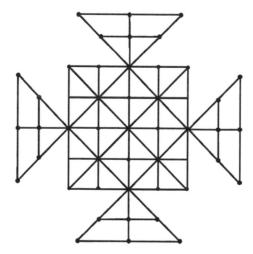

Figure 16. Cows and Leopards, also known as Sixteen Soldiers, is a two-player war game with an interestingly shaped board. The board shows that even in ancient times game designers understood the concept of discrete space and were experimenting with different patterns, shapes, and configurations.

has such elegance that one wonders if another game will ever match it, yet it is also so deep and complex that computers have a notoriously difficult time playing it. Created around 4,000 years ago, Go is still played competitively today, especially in Asian countries.

What's significant about Cows and Leopards (Figure 16) is that it's one of the first examples of an asymmetrical game—a game in which each force has different powers from the outset. This is in contrast to a game such as chess, which of course has symmetrical forces.

Chess is a game most of us are very familiar with, but not as many of us are as familiar with its history and all of the fantastic and interesting variants of chess that were created before, at the same time, and after the modern game evolved. I recommend looking into the history of chess, if for no other reason than to witness the course of evolution that it has taken.

Evolution as Inspiration

Very few ancient games were designed by one person. Most of them *evolved* over hundreds—or even thousands—of years. Because of this lengthy evolution, you can actually watch the collective design process of generations of human beings striving to find the right balance in the game. It's something every game designer should study. Part of this exploration should be finding copies of these ancient games and playing them. They can be inspiring in their elegance and simplicity, and beyond that, a lot of them are just great games. Good game design is timeless.

Sports in History

It is almost certainly true that sports predate any other kind of game. Due to their inherently physical and continuous, real-time nature, many ancient sports still hold up as games, not just contests (although there are many that are just contests too).

Each of the great ancient civilizations of the world has left behind evidence of its legacy of sports. Interestingly, many civilizations that had no contact with one another came up with similar types of activities. Most of these were contests like archery, running races, and other activities that are simple measures of physical abilities.

According to Steve Craig's *Sports and Games of the Ancients*, the top three sports that emerged everywhere around the world were running, wrestling, and archery. This was probably because all of them doubled as military and survival skills. Regardless, running and archery are certainly contests, since they require no decision making on the part of the players, and so we'll skip them and move directly to wrestling and other types of fighting.

Fighting

Any activity that involves fighting will almost certainly meet our definition of a game, because there are always choices in any kind of fight. One would have to come up with an absurd set of rules to turn fighting into a mere contest. For instance, a "fight" in which each player simply takes turns punching the other in the face until one is knocked over and loses would be a contest, not a game. But I can't think of any real-world examples of fight systems that are contests.

As I mentioned earlier, wrestling is *the* most common type of fighting game. We can see examples of it everywhere: from ancient Egypt (Figure 17), to feudal Japan, to various parts of North America. Even today, wrestling is an event in the Olympics. Because of its popularity, I should mention that professional wrestling of the World Wrestling Federation and World Championship Wrestling variety are not games or even contests. These "matches" *simulate* games through scripted performances.

The ancient equivalent of today's mixed martial arts (MMA) was probably the Greek pankration, a type of fighting with only two simple rules: no biting or eye gouging. (Still—those are rules!) As mentioned above, wrestling was and is found in cultures all over the world and has its own sets of rules (depending on the type of wrestling), and is usually highly regulated. Stick or sword fighting also was a worldwide phenomenon in ancient times, with local variants, restrictions, and goals.

Figure 17. A painting of ancient Egyptian wrestling maneuvers, found in tomb 15 at Beni Hassan (2000 BC).

In ancient Egypt, fishermen on rafts in the Nile River would fight each other with oars until one man was knocked into the water. We can see a modern equivalent in the television show *American Gladiators*, and not only is it still a game, but it can be very exciting to watch.

Ball Games

Ball games generally tend to fall into one of two categories: ball-and-stick games (such as lacrosse or hockey), and pure ball games (such as football or soccer).

One of the earliest sports, the ancient sport of hurling, has Gaelic origins that date back 3,000 years. Like Go, this is a game that is still being played despite its age. There are some aspects of hurling that should interest game designers. In many ways it looks like modern lacrosse, but in a design move that is very conscious of decision making, there are two goal areas in hurling. It's worth three points to get the ball past the goalie in a normal soccer-like goal, but there's also a second goal above that one that looks more like the uprights you'd see in football. Putting the ball between these uprights is easier, but it's only worth one point. The dual goals lend a very decision-oriented aspect to the game.

Even though most ancient sports (like most ancient board games) are contests and not games, I highly recommend that game designers spend some time researching them, as even contests can inspire interesting types of gameplay. For further reading on the topic, I recommend Steve Craig's *Sports and Games of the Ancients (Sports and Games Through History)*.

Playing Card Evolution

There are thousands of games that can be played with the modern deck of playing cards, and many variations on the classic 52-card deck that we all know. Playing cards first appeared in ancient China, eventually showed up in Europe in the 14th century, and have had a long, interesting path of design changes, including different numbers of cards in the deck, the types of suits, and the inclusion and types of face cards. The modern deck reflects almost a millennium of additions, removals, cultural references, spiritual symbolism, and changes based on pure game design (I'll go into more detail on card games in Chapter 5).

What's interesting, though, is that the evolution of playing cards was not driven entirely by game design: many cultural, political, and economic factors all influenced the way playing cards were used over time. For example, it's said that one of the reasons that aces are considered high (i.e., played as a high card rather than a 1) is because of social factors during the French Revolution. Playing with aces high was considered a small act of rebellion by the poor, and had very little to do with whether or not it made for better or more balanced gameplay.

The 20th Century

Some may be wondering why this book zips from ancient times all the way up to the 20th century—didn't anything significant happen between these two time periods? The answer is: not really.

As I mentioned in the introduction, human beings never had a colossal amount of free time until the 20th century. Without the kind of free time that those of us born in the last 100 years are used to, there simply isn't a huge demand for games. Modern games arrived with major technological advancements such as the automobile, telephone, electricity, and other products of the modern age.

Little Wars

Military exercises became common starting in the 1700s, with some military commanders creating great life-sized recreations of battles in order to study them. Sometime in the early 19th century, a Prussian general started to develop war games that involved small metal pieces on a large map and used rolls of the dice to help represent the possible outcomes of potential battles (see Max Boot's book, *War Made New*). All of it was purely for the sake of research, though—not for the sake of fun.

In 1913 H. G. Wells created a game called *Little Wars*, which is generally considered to be the first ever war game, a genre that lives on today (Figure 18). Of course, war games that simulate military strategy have always been around in the sense that generals and war strategists have set up simulations and maps to outline and test their battle strategies throughout history. But this was the first time someone created a simulation with fun as the primary intention (at least, the first time someone famous did it). And, as has been true throughout most of history, *Little Wars* was designed by a person whose primary job was not game designing. Until very recently, no one ever had that job because there was no commercial demand for it.

The war-game genre stayed somewhat dormant until the 1950s, when the first commercial board games began to be produced on a large scale. By the 1970s the genre was at its peak and had become extremely intricate and historically accurate, representing actual battles in the most realistic ways possible. Games like *Squad Leader* (released in 1977)

Figure 18. H. G. Wells playing *Little Wars*. Illustration first published in *Illustrated London News* (25 January 1913).

represented World War II battles, a favorite period for many war gamers. *Squad Leader* was followed later by a sequel of sorts called *Advanced Squad Leader*, which included countless scenarios and is still considered the most important war game today.

When looked at through our lens, war games are an interesting case. They are generally simulations, and as I mentioned in Chapter 1, simulations and games are not the same thing. But in the same way that a realistic soccer game is both a simulation and a game, a war game is also both a simulation and a game, because they are simulating something that happened to be a game in the first place. War *is* a game—it is a contest of *very* ambiguous decision making.

Pen and Paper RPGs

I go into more depth on pen and paper RPGs in Chapter 5, but there are a couple of historical notes I should make about these games first.

These games developed from the world of war games. In fact, the earliest version of what became *Dungeons & Dragons* was a war game called *Chainmail* (also created by Gary Gygax, one of *D&D*'s co-creators). Many consider *Dungeons & Dragons* to be the grandfather of video-game mechanisms, and for good reason: lots of early hobbyist computer games were inspired directly by the developers' experiences with *D&D*.

Of course, *D&D* is far from the only kid on the block in this genre. After *D&D*'s creation, tons of such systems began sprouting up. While early versions of *D&D* were heavily mechanical and more game-like, other RPGs coming out had a much greater focus on story and player interaction. It seemed that *D&D* wanted to stay on top of the latest craze, and it tended to absorb qualities of many popular games of the day with each new edition.

Today, a wider variety of these games are available than ever before due to the ease with which game materials are created and distributed on the Internet. Despite this, however, we see very little bleedthrough from non-*D&D* systems into mainstream video games (interestingly, many have claimed that *D&D* itself, now in its fourth edition, is heavily inspired by video games). The interesting world of pen and paper RPGs remains an untapped resource for new video game (and even board game) ideas.

Pinball

During the 20th century, we also saw a rise in various types of mechanical games, the most famous and important of which was pinball. Why is pinball important? Well, for one thing, we probably wouldn't have had the modern arcade without it. Many early video-game fans can recall being pulled into the arcade gaming scene initially by pinball machines.

Pinball has technically been around since the late 1800s, but it became a serious industry in the 1930s. By the 1950s we started to see a huge assortment of pinball machines with various themes, a trend that continued until the 1990s. At that time video games started to take over the spotlight and pinball seemed to collapse. At the time of this writing, only one manufacturer of pinball machines still exists, and it's extremely small.

So are pinball machines games? Well, it probably depends on the individual game to some extent, but I'd say in general the actual *meaningful choices* you get to make in pinball are extremely limited, if there are any at all. One thing I can say is that you certainly don't get to make any *interesting* decisions, so in terms of the game lens, pinball isn't a very good game. It looks much better when viewed as a contest. Entire books have been written on the subject of pinball alone, however, and I implore you to continue your research in this area on your own—going into more detail is beyond the scope (and the point) of this book.

The Promise of *SpaceWar!*

In the 1950s, with the invention of the transistor—possibly the most important invention in the history of electronics—the potential of making games with computers became viable. The very first game we have on record as being created using computers was called *OXO*, and it was simply a computer form of tic-tac-toe. I'm glossing over this because first, everyone already understands tic-tac-toe, and second, the use of the medium didn't allow for anything that couldn't have been done with a pencil and paper.

One of the first *original* games we have on record was one called *Tennis for Two*, created by physicist William Higinbotham at Brookhaven National Laboratory. Sometimes considered a precursor to the famous *Pong*, *Tennis for Two* actually included much more interesting gameplay despite having been created nearly 15 years earlier. *Tennis for Two* involved two players hitting a ball back and forth, and the ball not only responded to gravity but also to hitting the net (or even just grazing it!). Further, you could fake the other player out by waiting to hit the ball or by hitting it very quickly. There's quite a lot to this game, despite the fact that it was created on such primitive hardware—a 1950s computer and an oscilloscope display. The game physics (as they often do) serve to make decisions more ambiguous. We see physics used this way in sports too (most games have a ball throwing, tossing, or hitting mechanism for the same reason), and in most genres of real-time video games.

In 1961 Steve Russell, Martin Graetz, and Wayne Witaenem created *Spacewar!*, a one-on-one space combat game. The game was created on a PDP-1 computer at MIT. What's interesting about *Spacewar!* is just how well it holds up today. If one were to create a modernized version for Xbox or PS3 with a well-designed online hub for meeting players, a ranking system, and other such features that we've become accustomed to for online play, it could certainly achieve great popularity. The reason is that the gameplay, while extremely simple, is actually very deep and interesting.

To me, *Spacewar!* is the first true video game, in that it set out to create a completely original experience that could *only* have happened on computers. The gameplay involves a top-down game in which you and an opponent each pilot a rotating ship on a two-dimensional (2D) axis in real time. In the center of the screen there's a planet, which has gravity. Colliding with this planet will kill you, but a skilled player can use its gravity to help maneuver, dodge, and get a better position. You fire projectiles with one button and try to hit the other player to destroy him. You also can accelerate, but what's really interesting is that you also have a warp-speed function. This should only be used in emergencies, firstly because it teleports you to a random position (which is unsafe in and of itself), but also because with each use you increase the chances that using it will simply cause you to explode.

What's great about *Spacewar!* is that it really shows us the promise of digital games. Here we have a game that's calculating velocities and gravity, and randomizing things, all in real time. The outcome is that players are forced to make the sorts of decisions that they have never been forced to make before *Spacewar!* was invented. That is the promise of digital games.

How Far Have We Come?

Next I'm going to move into the formal video-game eras, but before I do, I want to stop for a moment and really consider a question: since *Spacewar!* was created, how far have we come?

Could it not be said that *Doom*—and therefore *Quake* and *Call of Duty* and *Battlefield*—are simply 3D versions of this same gameplay concept? The core gameplay of all of these is dodging projectiles and aiming your projectiles at the opponent. Of course, that doesn't make them all the same game, but it could be said that all of these games are *variants* of a game that was created in 1961, 50 years before the writing of this book.

As we'll review, and as many of you know, there are all sorts of genres of games in the modern world: platformers, real-time strategy, turn-based strategy, shooters, RPGs, etc. I think under duress, we could probably come up with about a dozen real genres. But using the lens that I am proposing—looking at games in terms of what kinds of decisions players need to make—does it not become clear that in a way, we've also only come up with a dozen or so *games*, with thousands and thousands of variants of each?

The standard for what innovation means in a digital game design in 2012 is very low. If you add any new gameplay features at all to an existing game, that's considered innovative. However, I can't help but feel that every game could—and possibly should—be fulfilling the promise of *Spacewar!*. That is to say, each game could be forcing us to make entirely new *kinds* of decisions.

It would be a mistake to claim that in the early days, everyone was innovating. But consider how different *Yars' Revenge* was from *Asteroids*, or *Pac-Man* was from *Donkey Kong*, or *Galaga* was from *Frogger*. Even games that were considered awful, such as the infamous *E. T. the Extra Terrestrial*, were very interesting and innovative (did you know that *E. T.*'s game map is the surface of a die?). Games like *Defender* and *Asteroids* presented entirely *new* verbs. The way you moved about, the objectives, and what "you" even were was interesting and new.

How many games can we say that about today? Very few. Recent examples probably include *Katamari Damacy* and *Desktop Dungeons*—beyond that, I can't think of any games from the 2000s that were actually games and were actually innovative with regard to their gameplay (*Portal* was innovative but it's a puzzle). Many are fooled into thinking games like *Swords & Sworcery* or *Bastion* are innovative because they have an interesting presentation. But that's all we've been getting: a continuous flow of new presentations of the same ideas.

I know that no medium has ever had an environment in which everyone was innovating all of the time, and that's not even what I want. The variants are good in that they are further exploration of what's possible with a given game. For instance, I do prefer *Team Fortress 2* to *SpaceWar!*. There should always be a balance, but in digital games we've been out of balance for a long time. We need more innovation from the ground up.

Video-Game Generations and Other Developments

Those who study the history of digital games tend to divide the existing history of games into generations. Indeed, console makers seem to all want to create new hardware with a similar rhythm of somewhere between five and ten years each. This is probably *entirely* the result of two things.

- Gamers have always erroneously believed that higher levels of technology equals better games.
- Moore's Law dictates that computing power will double approximately every two years.

These two factors (combined with some cultural factors surrounding video games) seem to have created a perfect storm for hardware manufacturers that has lasted nearly 40 years. Can it last forever? Already, brick-and-mortar retail businesses are hurting with digital downloads starting to change the way that people buy things. And some say Moore's Law may be broken in the next ten or twenty years. Let's take a look at the generations themselves and review what these individual eras meant for digital gaming.

The Beginning (First Generation, 1972–1977)

Those of us in the digital-games industry are familiar with the refrain "when new technology comes around, it's really going to change everything!" We're always waiting for the next generation of consoles and computers, because, well, just imagine what could be possible with more technology. But in truth, we haven't been doing much with the technology we already have—much of our potential is squandered. *Super Smash Brothers* came out in 1998 for the Nintendo 64, but it could have been created on the NES (so could *The Binding of Isaac*). We need ideas, not more technology.

But in the so-called first generation of games, I'd say that this catchphrase may actually have been true. These were really the first steps in digital interactive technology, the foundation of modern computing. So while these games were extremely important steps, very few that came out of this period have stood the test of time.

In 1972 Magnavox created the Odyssey, the world's first home digital-gaming system. These games are extremely rudimentary, with blank backgrounds, one or two squares, and a few lines onscreen. Many games are variants of air hockey, ping-pong, and table tennis. By the time Magnavox was finished producing new games for the Odyssey, only around 30 games had been created.

In 1975 William Crowther created *Adventure* (also known as *Colossal Cave*) for the PDP-10 computer. *Adventure* is a text-based game in which you type in various commands to navigate the world. This was arguably the first text-based adventure style of game, which would continue to evolve and lives on today as interactive fiction. Also in 1975, we see the first ever computer role-playing game with Don Daglow's *Dungeon*, a game that incorporated many elements from *Dungeons & Dragons*, which itself had only been created one year before, in 1974.

Very little truly creative work was being done in this first period—there were table-tennis type games, text-based choose-your-own-adventure style games, and simple get-out-of-a-maze applications. The name of the game in this era was, "hey, we can make something that is actually playable using computers." Like I said, other than a few outliers, most digital games to come out of this era aren't too helpful with respect to the development of our lens. I'd still recommend you research the period, but for the sake of developing our lens, we should move onto the second generation.

The Explosion (Second Generation, 1977−1983)

If the first generation was the sparking of a match, the second generation was a nuclear detonation. In the six years between 1977 and 1983, a massive, wide range of different kinds of games on all kinds of different hardware started to emerge, and quickly! This rapid development would of course lead to a rapid crash in 1983, but the conventions and styles of gameplay invented in this time period are still being followed today.

We can start with the arcade, which witnessed a massive boom and tremendous financial success, reaching annual revenues of eight billion dollars in quarters in its peak year of 1982. According to *Silicon Valley Fever* by Everett Rogers and Judith Larsen, in that year arcades took in more money than the film and music industries combined. Games such as *Asteroids*, *Pac-Man*, *Donkey Kong*, and *Space Invaders* began to show the world what digital games could really do.

To go in-depth and look at one example, let's talk about *Pac-Man*. Plenty has been said about *Pac-Man* before, but what's really important for us to note is how original *Pac-Man* really was. Sure, it's a maze game, and probably takes its roots from some of those Odyssey and early computer maze games. But the system of being chased by four ghost enemies, each with their own unique AI personality, coupled with the variable powers of the power pellets, which temporarily turn the game from an escape game to a predatory attempt to capture the hapless ghosts, is great. Further, even though it has a maze-like structure, it's not a maze in that there isn't a *start* or *finish* position. *Pac-Man*'s levels are about traversing the entire level. All of this is done, mind you, with no buttons at all—just a simple four-directional control stick.

It's obvious to most that *Pac-Man* is a classic, but what's important for us is that *Pac-Man* is uniquely a video game. Not only is *Pac-Man* original to video games, but it could not have been created using tools other than computers. It takes advantage of the fact that it is digital. Most of this credit should also go to the other arcade games of the time period, which scoured the possibility of digital games and searched for new possibilities as well.

During this period we also have an explosion in the home-console market, starting with the Fairchild VES and the Atari VCS/2600, and followed by the Intellivision and Colecovision systems. These systems were different from their predecessors in that they used cartridges for their games, which ultimately led to a much wider variety of available games than had been possible before.

One could write a whole book on the Atari 2600 alone. Its large and influential library of games spanned many genres and styles of play. It brought many arcade hits such as *Space Invaders* into the home, but Atari also produced some highly unique original games such as *Yar's Revenge* and one of my personal favorites, *Combat*. The Atari 2600 is also well-known for a top-down adventure game called *Adventure* (no relation to the text-based game from the first generation of video games).

Adventure is probably the most direct predecessor to something like *The Legend of Zelda*—a game in which you explore a large map, fight monsters, traverse mazes, and solve puzzles to succeed.

Home computing also became much more widespread during this time period, with all kinds of computers just aching to run some games. Computers like the Apple II, the Commodore 64, NEC's line of PC computers, Atari's 8-bit computers, MSX, and more. Further, a sort of "cottage-industry" of game developers started to take root, mailing out their programming instructions so that people could play their games on their home computers.

One of these developers was the now-famous game designer Richard Garriott, who created the first game in the *Ultima* series, *Akalabeth: World of Doom*, on an Apple II computer in 1980. This was a game that, like *Dungeon* before it, would strive to recreate (to some extent) the tabletop game *Dungeons & Dragons*. We'll see this theme popping up again and again in video games, continuing even to the present day.

I also have to mention *Rogue*, the inspirational dungeon-crawling game created in 1980 by Michael Toy and Glenn Wichmann on a Unix system at UC Santa Cruz (Figure 19). *Rogue* has since spawned over three decades of games inspired by it—so heavily inspired, and so proud of this fact, that they are referred to as roguelikes. They are notable for their randomly generated maps, turn-based combat and movement, and score-based play. My first commercial game, *100 Rogues*, was indeed a roguelike and many of its basic premises can be traced back to *Rogue*.

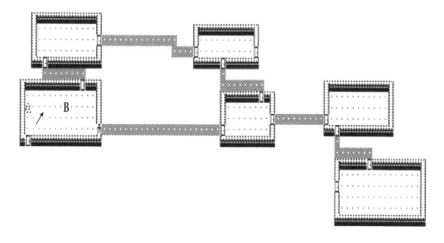

Level : 1   Hits : 13(14)   Str : 16(16) Gold : 0   Armor : 5   Guild Novice   Fast

Figure 19. The game *Rogue*.

During the second generation we also see the rise of companies that will remain big-time players in the industry (at least as of this writing, and probably well beyond it). Atari saw huge success, of course, but also we see the beginning of Activision, which was founded by some disgruntled Atari employees. In 1982, Activision created *Pitfall!*, which, along with Nintendo's *Donkey Kong*, was one of the first platformer games.

I really can't stress enough how important this time period was. During this time we saw more innovation in digital games than we will ever see again in such a short time. There was so much more to talk about, such as Nintendo's release of the Game & Watch line of electronic handheld games, or the first stirrings of online gameplay through dial-up bulletin boards. Things then were really great in the game industry, both for the players and the creators. According to Steven Kent's *The Ultimate History of Video Games: From Pong to Pokémon*, the golden age of video games started in 1978 with Taito's *Space Invaders* (a company that is still thriving). Unfortunately, such a rapid burst of success often has a downside, and the second generation certainly had its downside in the form of a massive crash in the North American video games industry in 1983 that the industry hadn't seen before or since.

Historians point to several factors that caused the crash, the best known of which was the failure of *E. T. the Extra-Terrestrial* and a poorly made port of *Pac-Man*, both produced for the popular Atari 2600. These two games were rushed to meet huge demand, and they were both expected to be massive hits due to the arcade success of *Pac-Man* and the cinematic success of the film *E. T.* When they finally came out they were pretty awful. *Pac-Man* didn't hold up at all next to its arcade version—and not just for trivial reasons, such as the graphics not being as nice, but because the gameplay was dramatically worse than that of the original game. Meanwhile *E. T.* was notoriously cryptic and weird to play, and left many players feeling as if it was unfinished. There's a famous story (legend, perhaps) about Atari burying thousands of unsold cartridges of *E. T.* in a New Mexican landfill.

But I don't think you can pin the crash on a couple of bad games (or even a couple of *really* bad games). I think it was a product of an extremely immature industry and culture getting much too big to support its own weight. I often feel that something similar could happen to us again, although perhaps to a less dramatic degree. The industry is in a different situation now: the standards for "bad" are much lower, and many of our journalism outlets are controlled, either directly or indirectly, by huge publishers. Great changes are coming, but who knows what form they will take. To quote Mark Twain, "History doesn't repeat itself, but it does rhyme."

Birth of the Fantasy Simulator (Third Generation, 1983–1995)

In 1983, just as the flames of the second generation had completely died out after the fallout in North America, a new spark was being ignited. That spark was Nintendo's world famous NES console. Soon the (now) barren landscape of the North American games industry would all belong to a new player: Nintendo. When the NES hit US shores it was almost like the industry had never crashed in the first place, and a new generation that would make even the second generation pale in comparison began. The industry picked up right where it had left off and exploded into new levels of success due to games that seemed to not only fulfill the promise of *SpaceWar!*, but possibly even exceed it. Of course, there were tons of hit games that were based on those of the previous generation and that were pure games, such as *Tetris*, *Gradius*, and hit sports games such as Nintendo's *Ice Hockey*. These kinds of games were still around, but something curious began to happen with this generation.

Although there were other 8-bit systems before it, as well as home computers that could perform similar technological feats, the NES introduced a massive audience to the idea that these new things—these digital applications that we've been calling games—could take a whole new direction. Games like *The Legend of Zelda*, *Metroid*, *Dragon Quest*, and *Final Fantasy* spoke to the potential for this kind of software to become a *fantasy simulator*. *Dungeons & Dragons* and computer RPGs, of course, were already ahead of the game in this respect. But those systems had a barrier of entry that was not acceptable to most people who weren't computer hobbyists or serious gaming types: the number of rules a player had to learn was simply too much for a young child or a parent who only had an hour or two each night to play games. Games like *The Legend of Zelda* or *Dragon Quest*, on the other hand, invited players from all walks of life to explore an exciting fantasy world. At least, that was the promise.

I was there, and I can attest to the level of excitement surrounding these games. I now wonder, though: was the excitement about the games themselves, or the promises that they made? *Dragon Quest* may represent the first time that players were asked to grind. And boy, did they ever. I myself can recall spending long nights "fighting" weak monster after weak monster to eventually get enough gold to buy that next sword.

By the time the NES went out of production in the mid-1990s, the modern video-game industry was fully formed. The kinds of games we played, the things we looked for in games, and the promises of games were all well established. While there were many brilliant innovations in this time period, I argue that we made a huge mistake at this point in history that we're still paying for today. That mistake was to not draw a line between digital games that met the design standards of traditional games (such as *Tetris* and *Spacewar!*), and games that really set out to do something very different (such as *Dragon Quest* and *Final Fantasy*). If we had drawn that line in 1990, I don't think I would have had to write this book today. It would have been a very different—and I mean better—world of digital games. So why didn't we draw that line? We simply didn't have the time. The medium was young and immature, and at the same time rocketing to stardom. Things were too good, too exciting, and too profitable for anyone to go about asking potentially *destructive* questions. And things remain that way, in large part, today.

The Perfect Storm (Fourth Generation, 1989–1999)

By the early 1990s, the video-game industry was healthy and strong, and actually began to settle down. Rivalries between companies formed, but what was particularly interesting about this era was that the modern language of digital games began to harden.

Improvements in technology, and hence potential gameplay possibilities, was very easy to see in moving from the first generation to the second. However, this had begun to slow a bit by the fourth generation. The technological advancements of the fourth generation largely allowed for things such as more colors onscreen, images with higher resolution, and higher sound fidelity. In other words, most of the games that came out during this period on the Super Nintendo could have worked on the original Nintendo with lower visual and audio fidelity. But that was not at all the case for games from the first and second generations: you couldn't have expressed the gameplay of *Pitfall!* on the Magnavox Odyssey, but you could have done *A Link To the Past* on the NES.

This was a source of concern for hardware manufacturers who, like most corporations, knew that defeat was only a mistake away from success. What resulted was a massive PR campaign that focused on the technological prowess of hardware. It's a basic tenet of advertising to talk up your weaknesses and turn them into strengths, so just as the newer hardware was becoming less useful, we were told that it was more important than ever. And we bought it. The games of the fourth generation were mostly higher resolution extensions of games from the previous generation. We began a sequel loop that we are still stuck in today, producing new *Final Fantasy* games, new *Mario* games, new *Zelda* games, and new *Dragon Quest* games. This situation sounds very familiar even to someone reading this book in 2010, 20 years later.

What's important about the fourth generation is that we begin to see games moving in a new direction during this period. Games started to get looser and easier, a development that some began to call *casual friendly*. This new direction was caused by (I might say, "allowed to happen because of") a perfect storm of two important and powerful factors:

- *The promise of the fantasy simulator.* Fantasy simulation meant that we didn't need to fuss about details such as whether or not the gameplay was interesting. *Final Fantasy's* gameplay was excessively repetitive and heavy on grinding; that is, most decisions were trivial. But we looked past that, because it wasn't about the gameplay. It was about simulating a fantasy, and this type of software became more and more common.

- *An expanding industry that needed to keep expanding.* Some would say an inherent problem (and certainly also an advantage) of a capitalistic economic world is that you can't be satisfied with the success of yesterday. Particularly in a market in which demand was exploding and the future was uncertain, companies needed to do better than they did last year, even if last year had been a banner year for the company. The feeling was that if a company rested on its laurels at all, the other guy was going to take over and it would only be a matter of time before the company went out of business.

Companies had to do everything in their power to stay alive. They had all seen many great teams collapse in 1983, and they weren't going to let that happen to them.

Of course, I personally wish that they had doubled down on their game design efforts. I remember saying in the mid-1990s that they were focusing too much on technology and story and all of that, and that they really just needed to focus on making fun games. So why didn't they? Well, they had no idea what made a fun game in any kind of solid, useful way. I myself didn't have any inkling about what it meant. If a company like Nintendo or Sega sincerely *had* wanted to just focus on making better games, what would that have meant? Hiring people who had made successful games? Instituting company guidelines for game design? What would those guidelines be based upon? What I'm getting at is that in 1995, there were no fundamentals established for game design.

We haven't formulated these fundamentals today, either. If a company today wanted to focus on better game design, the best it could do would be to hire people who *seem* good, based on whatever reasons seem like good reasons. My hope is that my book, and hopefully other books like it, will help establish design guidelines, so that the Nintendo of the 21st generation can focus on game design if it chooses to.

Game Shame in 3D (Fifth Generation, 1993–2002)

Much like the fourth generation, the fifth generation was essentially all about increases in graphical capability. By this point the largest and most successful companies had all but stopped creating interesting new kinds of gameplay. There were exceptions, of course, but for the most part, the fifth generation established the model for the entire next decade—and it was a 3D model.

The major innovation of the late 1990s was to make games look and sound like movies. Whether they were cartoonish games, such as *Mario 64* and *Banjo-Kazooie*; games influenced by anime, such as *Final Fantasy VII*; or the *Indiana Jones*–style adventures of games such as *Tomb Raider*, the goal clearly was to emulate cinema. In no game was this more obvious than in Konami's 1998 hit, *Metal Gear Solid*, a game that was fully voice-acted, had long cutscenes with careful cinematography, and featured a long, epic story.

Why did developers want to emulate anything else? Firstly, for the reason I stated before: if publishers and developers in 1995 had wanted to focus on game design, what would that have meant? Secondly, though, is because of a property I call *game shame*. Ultimately, publishers and developers were trying to make a living from spectacle—the wow factor, since that was something that was already well understood by that time. But why film? Well, in short, because we could. Not only was the film industry very successful at the time, but according to many at the time, film and video games were fundamentally pretty similar. But, as I've explained before, this is not the case at all. The properties that form the foundation of cinema are completely unrelated to the properties that make up the foundation of games—digital or otherwise. The argument went a little bit like, "Well, games have visuals, and movies have visuals. Games have music, and movies have music. Games have characters, and movies have characters. Sounds like a match to me!"

Of course, games don't need characters (*Tetris*). Games don't need music (or even sound) and actually, *games don't even need visuals*. There are plenty of games that are entirely verbal, such as the game telephone or the word game Ghost. One may argue that the addition of *video* to *game* requires that visuals be involved. Maybe, but I think that if someone released a digital game that didn't use the TV screen, with only sounds as its output, we would still call it a *video game*. Regardless—it's a game, and there is no need to draw a distinction between games that are digital and those that are not.

The emulation of movies during this era was really a tragic thing, particularly because this period was another era of exploding popularity for video games. I know many people whose first games were *Ocarina of Time*, *Final Fantasy VII*, or *Metal Gear Solid*. The world of 2012 is a world with a generation of people who think that the fifth generation was the era of the classics, and it shows.

Staying the Course (Sixth Generation, 1998–?)

Around the turn of the millennium we saw the creation of new consoles from Sony and Nintendo. We also saw the addition of a new player in the video-game market: Microsoft and its Xbox console. The sixth generation of video games began around this time, and besides the fact that Sega was out (its Dreamcast console did not do well after its North American launch in 1998) and Microsoft was in, there isn't much for us to talk about in this era. The reason is that this period represents a new low point in innovation in digital games. In some ways, the sixth generation is one in which the new consoles were the least needed yet, in that the number of possibilities was increased by the smallest amount. Most of the hit games were just updated versions of games that came before. The popular Xbox FPS *Halo: Combat Evolved* was a massive hit, but really brought very little new to the table. It was essentially introducing a new generation to mechanisms that PC FPS gamers had been playing for years. Sony's big hit at the time was *Grand Theft Auto III* by Rockstar Games (formerly DMA Design), which was a 3D version of its *Grand Theft Auto* franchise. And, while the beautiful visuals of Nintendo's latest *Zelda* incarnation, *The Wind Waker*, seemed to capture everyone's attention, in terms of gameplay it was much like the *Zelda* games before it.

Innovation on the Wrong Axis (Seventh Generation, 2004–Present)

By 2004, it seemed that even the console manufacturers understood that the industry was stagnating. The lead designer at Nintendo, Shigeru Miyamoto, has been quoted in saying that he personally found the fifth and sixth generations to be "very sad times." Correctly, he pointed out that games were becoming more and more about technology, and he said that he "didn't know who was designing his games anymore."

As it became more and more clear that simply increasing the graphical abilities of consoles was becoming less and less of a safe business plan, Nintendo started to look elsewhere for its seventh-generation console. The problem is, they looked in the wrong places. While the Xbox 360 and PlayStation 3 consoles pretty much stayed the course, the Nintendo Wii took a dramatically different approach. Essentially, the graphics capabilities of the Wii are the same as those of Nintendo's GameCube, but the selling point was the motion-sensitive controllers. Now you could fight skeletons in a dungeon by swinging your sword, get a real workout through a boxing match, etc. At least, that was the promise of the new hardware.

In practice, the Wii's new controller was lackluster; everyone had expected it to feel a little better than it actually did. It was slow to respond, sort of a pain to set up, somewhat inaccurate, and a bit finicky. Furthermore, the idea of one-to-one motion control just seemed to never happen. Even the latest *Zelda* game, *Skyward Sword*, which does use the swing motion and direction for your swing, doesn't allow for one-to-one motion.

There are many inherent problems with motion controls but actually there's a deeper issue: the input devices weren't the problem to begin with. These are game consoles we're talking about here. When it comes down to it, what people really want—whether they realize it or not—is good games. Of course, as I mentioned, neither Nintendo nor any of the other players had the intellectual tools available to know what this means. Without a philosophy of game design, game makers are forced to shoot in the dark.

Nevertheless, the Nintendo Wii was and continues to be extremely successful as of the time of this writing. I partially attribute this to the fact that while Nintendo may have been a bit misled in its quest for a new input device, it was absolutely correct in minimizing costs, thereby providing the system at a much lower price point than its competitors. Still, the larger problem has not been solved, and it seems that the future will be much less forgiving than the present.

Smartphones

Why, exactly, is the future going to be less forgiving for these powerhouse hardware manufacturers? One of the biggest threats is probably in your pocket while you're reading this. Early cellular telephones were primitive and manufacturers and service providers didn't offer much support when it came to games. The idea of playing games on a cell phone in the early 2000s was pure novelty, and very few games tried to be more than *Arkanoid* or *Tetris* clones.

The idea of smartphones—that is, cellular phones that allow the user to run programs and other operations as a desktop computer can—was not new at all in 2007. There had already been several companies making such devices as the Blackberry and the Palm. However, these devices were very specifically designed as business machines. Further, there wasn't much room for companies to make a lot of money, as people were more timid about making purchases online.

In 2007, however, Apple Corporation released its popular iPhone and introduced its App Store. Indeed, it had been Apple's own iTunes service that had helped many people feel comfortable about spending money online, and now iPhone owners were spending hundreds of millions of dollars on games and other applications in the App Store. This led to a massive wave of new developers working to create the next hit iPhone app. Indeed, my first commercial game, *100 Rogues*, was an iOS app—an opportunity I wouldn't have had without the iPhone. And I wasn't alone.

Google's Android OS and many other different types of hardware also started appearing on the market, each with its own app store. All of the sudden, Nintendo, Microsoft, Sony, THQ, Activision, and all the other big players had more competition—way more competition. Suddenly, anybody with the talent could put together a team and in a matter of months sell a game for one or two bucks in the App Store. Someone who, one year prior, the big companies could rely on to spend $49.99 on the latest *My Little Pony* Wii shovelware, was now buying three or four apps at one or two dollars each instead. This environment requires a very different sort of business plan.

Social Games

There have always been games that are social, but a new animal started to crop up in the 2000s that we began to call *social games*. (I'll get into the game design issues of social games in Chapter 4.) Social games are games wherein specific types of social player interaction are fundamental parts of the games. Blizzard's online multiplayer RPG *World of Warcraft* was a huge success, at one point reaching 11 million subscribers.

The problem for the big players was made even worse by the release of *Farmville* in 2009. *Farmville* was the first social game to reach a massive audience. That it reached this audience, by the way, was an outrageous, unprecedented success because *Farmville* was not made for a home console or even a smartphone. *Farmville* was available to everyone on Facebook, which almost everyone already uses. Not only that, but an inherent part of the game of *Farmville* is bragging about a new building you unlocked or a score threshold you reached—and, of course, telling your friends to come play. (*Farmville* awards players in-game bonuses for getting their friends to play. I think I'm showing considerable restraint by not commenting on this feature.) Unlike *World of Warcraft*, *Farmville* is not a subscription-based game. It's completely free to play, and makes money from small in-app purchases sometimes referred to as microtransactions.

Again, I'll get into social games more later (see Chapter 4's section, "Skinner Boxes"), but for now what's relevant is that social games were a major game-changer. The big console manufacturers went from having a near monopoly on the entire world of digital games to having to take serious inventory of their business plan. There is a very good chance that if they don't dramatically change the way they do business, one or two of them could be out of business within a decade. Indeed, during the seventh generation console manufacturers started to take some steps towards modernizing the way they do business by involving independent developers and by creating online platforms, such as the Xbox Live Arcade and PlayStation Network platforms.

The Alternate Reality of PC Games

I said that the big console manufacturers had a *near* monopoly, and the world of PC gaming has always been an exception to that rule. Not only does no single manufacturer have dominance over the PC hardware market, but independent developers—who now have *always* been a part of PC gaming—have an increasing amount of influence in software, largely due to modern forms of digital distribution.

As I mentioned, the first digital games and the early years of digital commercial games were largely home-brewed creations that were written to disks, zipped up in plastic bags, and mailed around the country. This independent hobbyist developer spirit has never gone away from PC gaming: in the 1990s, Sean O'Connor's game *Slay* made a lot of noise, and more recently we've had games such as *Spelunky* and *Desktop Dungeons* getting all kinds of attention. Software such as Game Maker and RPG Maker have evolved and become easier to use each year, allowing more and more people to express their game design ideas.

The culture of PC gamers has always been different—and I'd probably say *better*—than that of the console gamers. When I say better, I mean that they are a bit more in touch with that which matters about games. Why do I say this?

- The health of the indie development scene is one major factor in the culture, as it adds a bit of democracy, freedom, or just pure life to the PC gaming world. If you're tired of buying the same old stuff, you can always find some exotic new noncommercial masterpiece on some random developer's webpage.
- The more complex installation processes and input devices (mouse and keyboard) often invited developers to have slightly more intricate gameplay. While this is not necessarily a good thing, it did provide more freedom for innovation.

- *Patching* has always been a feature of PC games. This tuned players into some fundamental concepts of game design (e.g., balance), which console gamers may have taken for granted.
- I've often said that console games only have to be good enough to keep you playing while sitting on a super comfortable couch and portable games only have to be good enough to keep you playing while you're stuck on the bus. PC games, on the other hand, have to be good enough to keep you playing while sitting upright at a machine that could be doing a million other interesting things besides playing a game.
- As viewed through my lens, PC games tend to be better games. In particular, single-player games such as *Civilization*, *Master of Magic*, and *X-Com: UFO Defense* always have randomized content, whereas console games almost never do.

Of course, PC gamers have their own problems and biases, and by no means am I saying that PC gamers are completely enlightened about game design. But I think that they're generally *less* lost than the gamers in the console world. Regardless, if PC games or the culture surrounding them is any better, it's not better by a large enough margin to help the rest of digital gamers move towards a better future. If anything, the worlds of PC and console gamers have moved closer together in the past few years. While this has had some good effects, one side effect is that the culture of PC gaming has changed into something that looks a lot more like the modern console culture.

Other Notable Areas

There's a ton of stuff to cover, and not all of it fits neatly into a subject heading. In this section I'll quickly touch upon a few other areas.

Arcades

Video games had their first major successes with arcades, and these arcades would remain a major force in digital gaming until the late 1990s. Arcades were a gathering place, and between the social element and the fact that arcades generally had higher levels of graphics technology than home consoles, they were extremely *cool*. Most children of the 1970s or 1980s will always have nostalgic feelings about arcades.

Unfortunately, the temptation to exploit players for more quarters quickly became a major problem for these kinds of games. With the notable exception of fighting games (which were almost always competitive in nature), most arcade games were designed to be quarter-eaters, most often in the form of brawlers. Games like *Teenage Mutant Ninja Turtles*, *Golden Axe*, and *The Simpsons Arcade Game* were all extremely popular and extremely terrible games. Nonetheless, the cooperative nature of these games together with their popular themes was enough to keep us pumping in the quarters. I know *I* spent at least a hundred dollars of my parents' money as a kid playing *TMNT*.

The one great thing to come out of arcades, however, was the fighting-game community. This has lasted into the post-arcade era, and blossomed into a larger professional gaming world. Having a group of gamers who are so dedicated to playing games that they make it their profession is a very, very healthy thing for the world of games. If your game is exploitable, unbalanced, or otherwise broken, these people will find out. Better still, if your game is deep and interesting, these people will also find out.

Handheld Devices

Due to their hardware limitations, handheld devices have always been a sort of safe haven for games that focus on gameplay. Often these devices had hardware that was one to three generations behind the current generation of consoles, and because of that, the games often would focus more on having great gameplay, with less concern for fulfilling graphics expectations.

Sadly, very few of them actually ended up doing this, and so to this day we only have a handful of must-play portable games. One of the most successful handhelds of all time, the Nintendo Game Boy, was essentially a *Tetris* machine. I've played hundreds of the system's games, and I can't think of more than two or three other games that I would really recommend.

Nintendo's next major foray was the Game Boy Advance (GBA), which in large part was an *Advance Wars* machine. This was one of the most successful and most interesting games on the platform, particularly interesting because it showed that in the year 2001, a 2D pixel-art, turn-based war game could be successful with the right presentation.

Something must also be said about the Nintendo DS and DSi. While there are thousands of game titles for this system, I again have to say that *Advance Wars: Days of Ruin* (the second *Advance Wars* game for that system) is probably the reigning king. This game was notable in that it had online matchmaking and even had voice chat! Of course, computer games had already had voice chat for years, but this was probably the first example I had ever seen of a popular console game that featured it. *Advance Wars: Days of Ruin* is the closest to a PC game that I've ever seen on a console.

Most recently, Nintendo released the 3DS as part of the larger campaign apparently focusing on hardware gimmicks that was being used by all of the major console manufacturers. It had a very shaky launch, in part due to the 3D effect not working and/or causing headaches, and in part due to having very few games that anyone was excited about.

You may have noticed a heavy Nintendo bias in this section. That's because very few manufacturers have had any luck with their handhelds. Sega's Game Gear was a six-battery–eating monster with no interesting games, and Atari's Lynx was basically a bulkier Sega Game Gear. Bandai's WonderSwan didn't get a US release and essentially got murdered by the GBA. There are tons of other examples of failed handhelds out there—I encourage you to do some research on the topic.

I'd like to say that handhelds kept the major console companies and developers a bit more grounded, giving them license to do things that were, like *Advance Wars*, turn-based or low-res or lightweight and snappy (as opposed to clunky with huge load-screens, multiple cutscenes, and animations that can't be skipped). Unfortunately, this wasn't the case. It's not clear that the success of various handheld games affected the way that the other console developers created games. It seems as though handheld games have always been considered in their own little bubble, along the lines of "*Advance Wars* is good, *for a handheld game*."

Renaissance of the Designer Board Game

Since the 1970s, board games have been *a thing* for the Germans. I'm not completely sure why this is, but for some reason, they took to game design very well. In 1995, the world (and particularly the United States) noticed when Klaus Teuber released his hit game, *The Settlers of Catan*. Many hits followed, and we now have a world of designer board games that is rapidly expanding in a way no one could have predicted.

As I mentioned before, these games are called *designer* board games because they feature the game designer's name right on the front of the box. I also think that they could be called *designer* because when you compare them to almost any other kind of game that has ever been made, these really seem *designed*. There is a methodology, an understanding of the true nature of gameplay in these games.

My personal understanding of games would not be where it was today if it weren't for designer board games like *Through the Desert, Puerto Rico, The Resistance*, and *Yomi*. I regret not having discovered them sooner, and I think that many people will feel the same way when they discover them too. I talk at length about these games in Chapter 5.

Looking Back

As I stated earlier in the chapter, we've only recently entered this new period in which there's actually a commercial demand for games. And with all of these new platforms and possibilities, the potential has never been greater—yet, year after year, it seems like we're sort of treading water. Most digital games that are good almost seem like they were good by accident. Developers are shooting in the dark.

Worse, we're now in a position where a large percentage of our games are merely reskinned clones of one another. We have the RPG, the third-person action game, the FPS, a few RTS games, the puzzle-platformers, and games based on sports. That list makes up 90% of the new releases in any given year. Worse, a system of check-box game design has been bringing us closer and closer to all games being the same thing: a first- or third-person action game with cover, health regeneration, quicktime events, and RPG elements. Still worse, the modern gamer has come to expect new games to be like Advent calendars: look in all of the windows, consume the content, and then you're left with garbage to throw away.

So how did we get to where we are? Well, actually, *we never really changed.* We are still mostly the same regarding games as we were thousands of years ago, it's just that we're suddenly being asked to produce hundreds of games each year (when we used to need maybe one new game every hundred years!). We're throwing mechanisms against a wall and hoping some of them stick. We're thinking up themes and building systems that express those themes. In short, even if we're paid to design games, we aren't thinking like game designers.

Game Shame and Immersion

First, we have to develop a serious respect for games. It's not surprising that no one has a good understanding of what games are and how they work, since they're considered frivolous activities only suited for children. Video games are a joke in our culture.

For about six or seven years, I played drums in a band that played covers of songs from famous video games. Our repertoire was made up of the soundtracks from *Super Mario Brothers, Mega Man, Castlevania,* and *Sonic the Hedgehog,* just to name a few. And we would do shows—most of which were for an audience of hardcore gamers, oftentimes at gaming tournaments or other gaming events. And there was a recurring theme to how people responded.

I should mention first that the reaction was positive—it's not that people didn't like the way we played. The weird thing was that everyone seemed to assume that the idea of people playing video-game music was inherently funny. When I told people about the group they sometimes would burst into singing the theme from *Super Mario Brothers*, grinning and giggling as they went. And let me be very clear—although it doesn't come through in the writing of this book, I am a person who is very much in touch with comedy. And yet, I couldn't see where exactly the humor was in humming the notes of a video-game song. But eventually I figured it out. The problem was that the joke relies on a premise that I believe to be completely untrue. The joke is, video games are inherently stupid.

Once you realize that this is the fundamental underlying narrative of the culture with respect to video games, so much else starts to make sense. This is why we try to make our games look like movies—because movies aren't inherently stupid! This is why we have a movement called art games—because art isn't inherently stupid! This is why we focus on technology. This is why we focus on immersion. This is why we believe today that games can only be a means to the end of an eventual virtual reality fantasy simulator. None of those things are inherently stupid, but games are.

I refer to this condition as *game shame*. I can't say for certain when it started, although it's likely to have existed well before the advent of digital games. Was there ever a time when game designers were put on the same pedestal as musicians, architects, or painters? I don't know, but I know that in order to move forward we have to start respecting the medium.

Preserving History

Want harder proof that game shame is a phenomenon? Look up the term *abandonware* on the Internet. Abandonware is basically *all* digital games created before the year 2000. (The reference year, by the way, is constantly moving forward in time—as of 2012, it may actually be more like 2003 or so. The cutoff date tends to be the current year minus about 10–15 years.) Regardless, the point is that you literally *cannot* buy a copy of most of the digital games ever created. The company doesn't produce them anymore, if they even exist. Not only that, but even if they do exist, some publisher owns the rights to that game. So not only can you not buy it, but you can't legally download it from anywhere, either.

Luckily, many people took up the mantle of preserving video-game history by creating large databases of these abandoned titles. Some of these sites, such as Home of the Underdogs and Abandonia, have extensive articles written about the games, manuals, and other accessories in addition to making the full games themselves available for download. Although these sites dance on legally murky grounds (at best), they are doing us all a huge service by being, in essence, a makeshift library for games.

I propose that we should preserve games in publicly funded libraries. Of course, what that will mean in the post-paper future has yet to be seen. Either way, whatever institution is protecting the works of Mary Shelley and Edgar Allen Poe should also be preserving the works of Richard Garriott and Tomohiro Nishikado.

Why Some Stuff Stuck

Why is Go and chess still played everywhere in the world, while some other games that we have a record of aren't? What are the properties of games that became, and stayed successful? What gives a game staying power?

Of course, I believe that I have distilled the answer via the lens featured in this book. But I encourage you to find your own answers.

4

Through the Lens: Video Games

I n this chapter, we'll be taking a look at various genres of games, in order to develop the accuracy of our lens further. Genres are always a bit of a tricky thing to nail down exactly, and there are all kinds of differing opinions out there about which games belong in which genres. Some people don't consider *Super Smash Brothers* a fighting game. There's great disagreement about whether or not *Diablo* is a role-playing game. Further, there are some "genres" that are simply too broad to have any meaning, such as action, or to a lesser extent, strategy. For this reason, I'll be defining the genres as I see fit. But don't worry—I'll also include an explanation illustrating what I mean by these genre names.

I'll have suggestions for designers regarding these games as well— what pitfalls to avoid, and how we can look outside the box of these genres to create truly new gameplay experiences. If you're designing a game that even sort of fits into one of these genres, reading the relevant section should be helpful. If you're looking for ideas, there are many here, ready to be pursued!

Problems Common to Most Genres

When I began writing this chapter, I jumped right into the first section on individual genres: "Brawlers." I wrote it, and then when I started to move on to the second section (third-person action), I realized that many of the problems that applied to brawlers would be appearing in the second section too. In fact, I realized that there are key problems that are common to almost every video-game genre. These are things we've touched upon in the book before, but since they're relevant to this section, let's get them out of the way first.

The Accidental Puzzle

Intentionally or not, any single-player game that doesn't have random elements (randomized level layouts, randomized monsters, etc.) is going to degenerate into a puzzle or a contest. On your second play of *Castlevania* or *Mega Man*, you've already memorized some of the game. These games require execution, and so they become contests. An adventure game such as *Secret of Monkey Island* or a turn-based game such as *Advance Wars* do not have execution elements, so they become memorization puzzles very quickly.

For this reason, I advocate that all single-player video games have random elements. Games should be testing players' skill, not their memory. Imagine that you wanted to examine children's addition skills: there are two possible ways that you could test them.

- *Method A*. Give the children the same math test, with handpicked questions, ten times over.
- *Method B*. Give the children ten unique math tests, each with their own randomly generated problems.

With both methods, after the children complete one test, you would grade it and give it back to them. Obviously Method A has the advantage of having better quality control: there won't be any extremely easy questions in the one with handpicked questions. However, after the first couple of repetitions the children will begin to memorize the answers, and after doing four or five tests, they won't be doing mathematics at all—they'll only be repeating what they've memorized as the correct answers.

This is what happens with single-player games that have no random content. Players simply begin to memorize. Yes, a random generator may produce less interesting results, but even after the very first play, a handcrafted level becomes less interesting than the least interesting random level or game situation.

The Fantasy Simulator

As I explained in Chapter 2, at some point we collectively decided that video games would be primarily fantasy simulators. Of course, the task of simulating a fantasy and creating a game are two very different tasks with very different requirements. Fantasies don't have to be balanced, and fantasies don't have to be elegant. More actually may be *better* in fantasies. Fantasy simulators don't need to have meaningful, ambiguous decision making.

So, if we are assuming that hit video games like *StarCraft*, *Metal Gear Solid*, and *Mass Effect* are supposed to be games (which, granted, is a big assumption—if the designers of these games saw my definition of *game* they may well say that they weren't going for that), they don't do so well. If they're supposed to be fantasy simulators, they do a bit better. I believe, though, that this decision was never consciously made, as most people don't distinguish between the two. If they had made that decision, they'd be making better games, or better fantasy simulators.

More Is Not Better

The close relationship between video games, technology, and American consumerism has led video-game design to become a craft of excess. The backs of the boxes brag about features like 200 spells, 70 guns, or 50 characters. This obsession with more, or bigger, or higher levels of technology has led game designs that are more and more watered-down. The concept of elegance—a fundamental aspect of not just game design, but design itself—is totally alien in the video-game world. The question is not, how can we express our idea in as few mechanisms as possible? The question is, how many mechanisms can we afford to cram into this thing? The result is watered-down design, choices that are false and uninteresting, and games that are impossible to balance.

Tied Down by Story

In Chapter 2's section titled "Games and Story," I explained why the relationship between games and story is harmful—to games. This problem is related to a game being driven by theme, but it's a very special part of that flaw. Story is special in that it's so completely unnecessary and easy to avoid—in fact, it actually takes a lot of work to add a story to a game. A story-based game is more expensive to make and inherently has less replay value. That's what I call inefficiency.

3D and Camera Controls

We live in a world of 2D joysticks (D-pads, thumbsticks, mice, and arrow keys) and 2D output devices (computer monitors and television screens). If you can, make your gameplay 2D. So many problems come from trying to emulate 3D space on a 2D screen with a 2D controller. For one thing, players have no depth perception; it's like they're running around trying to gauge precise jumping distances with only one eye.

Another huge problem that's often (but not always) an issue with 3D is that of camera controls. If your players have to stop playing every few seconds (or even every few minutes) to fix the camera so that they can see the information necessary to play the game, you have, on some level, failed to do your job as a game designer. This totally breaks up the experience and is annoying and completely unneeded.

Focus on Metagame

So-called RPG elements are now expected to appear in every single genre of game, no matter how inappropriate they may be. In this context, RPG elements refer to the *metagame*—another game that wraps itself around the game in question. This other game is very rarely interesting, and almost always something of an afterthought. The metagame is usually a very bad and broken game when taken on its own.

The simplest form of a metagame would be something like a record of your wins and losses, or a high-score board. But usually when people refer to the metagame, they mean things such as unlockables or customizable features. The problem with metagames is that they almost always damage the games they surround. The first way they do this is by reducing the importance of decisions in the original game. For example, let's say you have a simple one-on-one fighting game. When you add several types of metagame features on top of that, those metagame decisions you make are going to have an impact on the game itself, and therefore reduce the impact of the in-game decisions.

Metagames also often create motivations that conflict with the inherent motivations of the core game. The developer, Valve, experienced this in *Team Fortress 2* with its unlockable weapons metagame element. It tried many different ways to allow players to unlock weapons, which were usually based on getting a certain number of achievements. But it was clear that if the metagame feature was tied to anything—including achievements—people would stop playing correctly. And they did! Players essentially began to grind the achievements instead of playing normally.

The most common defense of a large number of metagame elements is that it provides variety. However, metagames always make games much harder to balance, because every element of the metagame essentially multiplies the number of interactions in the system. Therefore, dominant strategies are more likely to emerge, reducing variety. Not only that, but each element's flavor, or difference from the other elements, is diminished because of the overlaid metagame mechanisms. In a game with a significant metagame element, there ends up being *less* variety because everything becomes a big, blurry wash of mechanisms that lack identity.

Three-dimensional games, at least, have little choice but to include camera controls—but there are actually 2D games that allow players to move the camera around for some reason. Real-time strategy (RTS) games and computer role-playing games (CRPGs)—such as *Fallout*—are two examples that come to mind. Make the level small enough that I don't have to scroll around. In any game that has a mini-map, the mini-map effectively becomes the game, because it actually makes sense. Just make the full screen the mini-map. Come up with creative solutions; don't make me do your work for you when I'm playing the game.

I should mention there are two genres that provide notable exceptions to the avoid-3D rule, and that's because these games, by their nature, sort of need to be in 3D. These two genres are first-person shooter games and racing games. In both games, however, the camera usually is not something separate from gameplay that you have to worry about. In both FPS games and racing games, the camera position generally is dependent on where you're actually pointing in the game. If a game ties the camera to a fundamental element of gameplay, then it's OK. I'll touch on this more when we get to these genres.

Brawlers

- Examples: *Golden Axe, Final Fight, River City Ransom, Double Dragon*

Brawlers, as they were called in the coin-op arcade days, are side-scrolling games wherein you control a character (usually a burly muscleman or martial artist) who has to fight his or her way through several levels filled with bad guys. They were extremely popular in the 1990s, particularly in arcades where they made for excellent quarter-eaters, but we don't see too many of them these days (Xbox Live Arcade had a *Scott Pilgrim* game and there was also *Castle Crashers* on that same platform, so it seems that the XBLA platform is to some extent a holdout in this genre). Generally, third-person action games are taking their place these days.

Brawlers tend to have health bars, lives, and continues, and they also tend to allow for 3D movement onscreen, despite their side-scrolling nature. The term *Z-order* refers to how deep the player's plane (the playing field) is on the screen in a brawler (or other games that simulate depth in a similar way).

There are several serious problems with brawlers, but like almost every other genre, their problems tend to stem from the same underlying problem, which is that their designs are driven by the theme and not the mechanics.

The first serious problem with brawlers is the issue of Z-order. Anyone who has played a brawler knows that in almost any of these games your enemies can never attack you as long as you stay out of their Z-orders. That's a dominant strategy in just about any brawler I've ever played. Unfortunately, it's also a source of confusion, because it's not always clear if you're precisely on the Z-order of opponents or not. This issue of Z-order plays into another issue that comes up with third-person games as well.

Also, most brawlers are not randomized; they're totally linear. This means that after a few plays, players will begin to memorize large chunks of the early game. If players have already memorized optimal moves, the game is *dead*.

The games are also, frankly, gigantic messes that are held together *only* by their themes (and even those have holes in them). You have health bars, which already is questionable (I'll get to this in the section on fighting games), but then you also have lives, each of which, when consumed, fills your health bar. On top of that, though, you also usually have *continues*, which fill your *lives* counter when consumed. You can see that these three mechanisms are essentially all expressing the same element, but obscuring it at the same time by putting it into three different denominations.

And I hope that I don't need to explain how it's insane, illogical, and possibly even immoral to allow a player to essentially buy more in-game strength by putting more quarters into a machine, but the recent rise of social games and in-app purchases makes me wonder. If you have 100 dollars to spend on any coin-op brawler, it doesn't stand a chance. This is in stark contrast to the older model of using credits, which allowed more plays of a game, but still started you from the beginning when you died. By the end of the arcade era, credit meant "continue," or "you now get a free health bar multiplied by a set of lives."

Some suggestions for building better brawlers follow.

Ignore the Theme for Now

Punching bad guys is thematic—mechanically, though, what is your game about? Is it about controlling space on the screen? Is it about synergies between tactical moves? Is it about managing resources? Maybe your game is about predicting randomized patterns of enemy movement. The point is, start out asking the right questions.

Ditch Z-Order

The Z-order idea almost certainly needs to be ditched. It simply doesn't make any sense. Why can I move up and down, but not attack up or down? And it will never be visually clear, because we're talking about continuous space on a band of pixels with no depth perception. One possible solution is to do something like what *Mega Man Battle Network* did and create a discrete grid of squares, rather than continuous space, that you move around on.

Instead, ask yourself if that 3D movement is really necessary. Many of these games would be much better off with a purely top-down scheme, but perhaps making it totally 2D (side-scrolling) would also work. Maybe your game is all about players jumping off of each other's heads. Think of *Super Smash Brothers* and how rich that is purely on a 2D plane: that's what can happen when you start a design from the ground up.

Consider Semi-Cooperative and Score-Based

Consider developing a semi-cooperative brawler. Maybe all the players can lose, but only one player can win. Maybe each player has his or her own way of winning that is unique. Look to board games like *Chaos in the Old World* or *Battlestar Galactica*'s *Pegasus* expansion for great examples of how this can be done.

Also, consider taking a more roguelike approach in which the goal is to beat a previous high score (it won't work unless you randomize the game, but as I've made clear, you must do this anyway). Score-based games are great because the challenges are always renewable.

Think Nonlinear

In the late 1980s and early 1990s, a very cool company called Technos Japan (made famous for their *Double Dragon* games) was experimenting with some very interesting new breeds of brawlers. In the United States, we got its *River City Ransom*, which combines elements of the RPG genre with the brawler genre and allows players to explore a map freely and somewhat out of order. Lucky Famicom console owners got to play *Downtown Special: Kunio-kun no Jidaigeki dayo Zen'in Shūgō*, an extremely innovative, nonlinear brawler with an overmap and enemy lords to chase down. The point is, making these games nonlinear can really work, and Technos gives us numerous examples of how it can be done.

Punish Button Mashing

If players are doing alright while button mashing, you've failed. Button mashing involves, by definition, players not making decisions. It's not hard to solve the problem of button mashing, if you actually realize what a failure in game design it is.

One solution is to have attacks have an element of commitment: for instance, building in a delay after a punch is thrown. If you miss, you're left vulnerable for a few seconds. Another very easy solution is to eliminate *combos*. You might think it's illegal to make a brawler that *doesn't* have the punch-punch-knee combo, but it's not. If anything, it should be illegal to add that combo to any more games. These combos add almost nothing to the actual possibility space of the game and are only there for thematic reasons. They should be ditched.

3D Third-Person Action

- Examples: *God of War, Devil May Cry, Dynasty Warriors, X-Men Legends, Tomb Raider*

Of course, that side-scrolling brawler stuff doesn't look *anything* like a movie, so it's mostly a thing of the past these days. Now, the camera is placed *behind* the main character, but other than that the gameplay is largely the same: get close to enemies, and mash buttons to kill them. One becomes good at the game by memorizing AI patterns, monster spawn positions, and level geometry.

Usually, 3D third-person action games have the player using a melee weapon, but there are some exceptions, such as *Tomb Raider*. However, I still classify *Tomb Raider* in this category because the gameplay is essentially the same as *God of War*: your weapon simply has more reach since it's a pistol instead of a fiery god–whip.

Third-person action games are the modern extension of the brawler, and so they share some of the same problems. The 3D feature brings a lot of new possibilities, though—most of them problems in disguise. For that reason, and because of the sheer magnitude of popular 3D third-person action video games being produced, they get their own genre.

Camera Buttons

As I said at the beginning of the chapter, if your game has camera buttons, you've failed. And almost all 3D third-person action games have camera buttons. What this means for players is that every once in a while (or in extreme cases, frequently), they have to stop playing and take measures to correct the camera angle. Let's do away with this.

How different would the gameplay of *God of War* be, if it were totally top-down? Or if perhaps it had a three-quarters-perspective view, as in *Diablo* or *Final Fantasy Tactics*? Tragic as it is, the gameplay of *God of War* is 2D.

Watered-Down Features

So many of these games have a number of moves that have nearly identical effects, which are only differentiated by visual representations. They often have features such as jumping and ducking that have, at best, shaky relationships to the core mechanisms of the game. Even combat moves generally have very little gameplay identity to distinguish them from each other. If you have two sword swings that both do the same thing, one of them should be cut out. And as with brawlers, combos need to go unless performing combos requires strategic planning and has an element of ambiguity.

Real-Time Strategy

- Examples: *StarCraft*, *Warcraft* (not *World of Warcraft*), *Command & Conquer*, *Total War*

The genre of real-time strategy games is fundamentally flawed. First, let me say that I essentially dedicated my teens *and* my twenties in large part to RTS games. I first started with *Warcraft II*, which I played on my dial-up modem online through an early online gateway called Kali. I then played *StarCraft* and *Red Alert* intensely for years. My attachment to the genre peaked, though, with *Warcraft III*. I not only played it, but also did audio commentaries on replays for it, and even wrote articles about the game and posted them online. (In fact, that's how I got started writing about video games. My first article, which I wrote for a site called WCReplays, was translated into several languages and got hundreds of thousands of views.) When *StarCraft II* came out in 2010 I played it pretty intensely for about six months, and then it clicked—I realized then that I would never play this kind of real-time strategy game ever again. That's not because these games are all bad games; in fact, they have a huge leg up on many other video-game genres because they stress multiplayer, competitive play. But with this said, I also must observe that strategy games should not be in real time.

The term *strategy games* is a bit blurry, since all games involve strategy. But usually when we say *strategy games*, we're actually speaking loosely about war games. War games are about moving a number of units around a map to overwhelm an opponent. Now, real-time strategy games have two totally distinct elements to them: your strategy (which is a conceptual plan that you have chosen to undertake) and your execution (which is essentially you telling the system what it is you want to do).

In games, execution—meaning actually communicating what you have chosen to do—should never be what's difficult. Execution should be a foregone conclusion for any strategy. Strategy—making choices about what you are going to do—should be what's difficult. Now let me clarify: some strategies are difficult to execute because they require that a lot of things go your way. For instance, playing as a spell-casting character in *Dungeon Crawl: Stone Soup* is difficult, but not because inputting individual moves is hard. If you've chosen to cast a spell, doing so is trivial. Knowing whether or not to cast that spell is not trivial.

In RTS games, you can choose a winning strategy but still lose if your execution doesn't meet the dexterity requirements—which get pretty intense in these games. There's even an acronym for how fast people issue commands in the games: APM, which stands for *actions per minute*. This isn't to say that good strategies are not ever rewarded, but you have to meet a dexterity requirement first.

As I mentioned in Chapter 3, games have a certain resolution of input, which dictates how much information can be sent to the system in a given moment. In RTS games, the resolution of input is so astronomically high that it's effectively infinite. You can change the screen position, and you can move hundreds of units to almost any continuous position. You can issue some six to eight other commands to various units, depending on what they are, at the same time. Many units have abilities that can be cast in continuous space. You usually also have buildings that all have various abilities and need to be clicked on repeatedly throughout the course of a game.

In short, there is basically no real cap on how many commands you can issue in a second. This means that the importance of strategic decision making will be diminished and the importance of execution and managing the user interface will be huge.

Is Scrolling Necessary?

Why does your RTS game need scrolling? Must the maps be so big? If so, why? If the reasons behind these elements are thematic, I've already explained why this is not good design in Chapters 1 and 3. Games are mechanical, and the theme is a layer placed on top. Building a game around a theme is as smart as building a castle on a swamp.

Perhaps you have some nonthematic, mechanical justification for having huge maps. Well, if that's the case, is it possible to make the units very tiny, so that the map can fit on one screen? Either that, or you could tie the scrolling to the gameplay. Maybe things can only happen in an area when your screen is focused on it. Maybe you move units by scrolling the camera (the camera is always centered on one unit, for instance). What you should be trying to avoid at all costs is the player having to manage some system that has no connection to gameplay, which is what scrolling usually is.

Watch the Inherent Complexity

RTS games are inherently extremely complicated. They feature ridiculous statistics like armor ratings that reduce incoming damage by absurd factors (such as 0.625% per point); units (and there are usually close to 100 types) that have individual, arbitrary amounts of health, damage, armor (often with several different *armor types*), and speed; several different resource costs; build time; requirements; and a list of abilities (each of which have their *own* stats like damage, cooldown, mana cost, blast patterns, and so on). The amount of arbitrary information a player has to learn to even play these games is totally, completely outrageous, and a mark of horribly weak game design.

A game as complicated as *StarCraft* will *never* be balanced—there are simply too many factors and interactions at work. Even if *StarCraft* were 100 times simpler, we would still be talking about a ridiculously astronomical number of meaningful interactions (especially given that the game takes place on a continuous space). And not only is there too much stuff, but we now have an expectation of perpetually adding more and more stuff into the system for years on end.

In short, be careful of the amount of content in these games. Remember, lasting emergent complexity that comes out of a well-designed system is the ideal.

Consider Symmetrical

By convention, most RTS games feature asymmetrical forces of some sort. There's nothing inherently wrong with this, but creating an RTS with *symmetrical* forces is certainly worth considering. In fact, you should be extremely careful when adding asymmetry to your game, since you're starting on a task that will increase in size exponentially and quickly become impossible to balance (see Chapter 3).

What Is Your Game Really About?

RTS games have a habit of being about many things: base building, army building (choosing counters), tactics, and resource management. I think that the way forward for the genre is to choose one of these and make it the core mechanism of the game. There can still be other mechanisms in the game, so long as they are in direct support of the core mechanism.

The obvious counterargument to this proposition might be something like, "a game about resource management would be boring." Firstly,

I don't think that that has to be true at all. There are plenty of fantastic games that are just about managing resources—you just need to have some interesting things you can do with the resources (such as bidding on them or trading them). But secondly, if you know that something is uninteresting and the reason that it's uninteresting, then why is it in the game? Things that are trivial or uninteresting can probably be removed safely.

Consider Turn-Based

A strategy game is ostensibly a game about making *strategic decisions*. It is not a game about timing, or precision. Of course, you're welcome to have an execution element in your strategy game, but I can't see a good reason to do so. Although a real-time game doesn't necessarily have to include an execution element (or if there is, it could be negligible), it's worth considering making your strategy game turn-based instead.

Turn-Based Strategy

- Examples: *Advance Wars, Fantasy General, Final Fantasy Tactics, X-Com: UFO Defense, Civilization*

The story of turn-based video games is a strange and sad one. Even though video games take inspiration largely from board games and *D&D*—both of which are turn-based—video games have largely made *turn-based* into a bad word over the last decade or so. This is probably a byproduct of the video-games-as-fantasy-simulators movement and the idea that video games should be as cinematic as possible. Turn-based games are anything but cinematic—they *dare* to be shameless about the fact that they are games.

Early on we had many turn-based games, particularly in the world of PC gaming. The developer and publisher SSI was well-known for the *Panzer General* series of computerized war games (as well as the fantastic, more game-like and less simulation-like high fantasy spin-off, *Fantasy General*). We also had a big movement featuring so-called 4X games (which stands for explore, expand, exploit, and exterminate) like Sid Meier's *Civilization, Master of Orion, Master of Magic, Heroes of Might & Magic*, and later, *Age of Wonders*. The options were more limited for consoles, and many of the most popular turn-based games were JRPGs (Japanese role-playing games), such as *Final Fantasy* or *Pokémon* (we'll get into those more in the upcoming section on RPGs). The gameplay of 4X games was really nothing special, and they may have had something to do with the decline of turn-based digital games.

Waiting for Animations

In a turn-based game, making the player wait for animations to play out is a problem. This is very low-hanging fruit that can be fixed easily if you understand the issue, so I've put it first. You see, in a real-time game, animations have a gameplay meaning. No one is waiting around while an animation plays out in *Street Fighter* or *Call of Duty*. In real-time games, the time it takes to perform a particular move is a crucially important part of how that action is balanced in the game. A move is commonly very strong but balanced out by a long cooldown or reload—a delay.

In turn-based games, however, animation does not have any gameplay meaning: it just makes the player wait while a little movie plays out. This may have reached a peak of awfulness for *Final Fantasy XIII*, in which the various spells you cast resulted in cutscenes that lasted minutes and couldn't be skipped. But even in smaller doses, this problem gets bad quickly. Let's say there's a one-second animation every time you issue a command. If you're pretty good at a game, you might play rather quickly. Let's say you issue a command every 2 seconds. That means that if you play a game for one hour, you've actually spent 40 minutes playing (issuing commands) and 20 minutes sitting and waiting for animations to play out. Those seconds add up!

Here's what you, the game designer, must do: if you need to have animation at all (and there's often no good reason), decouple it from the player's input. That way, if an animation isn't finished and a player performs another action the animation will either be interrupted or will continue to play out while allowing the player to do something else. Asynchronous animation might be another way of putting it.

Elegance Still Counts

It seems that many designers *spend* every drop of UI improvement they get from making games turn-based on making many of these games as complicated as possible. *Heroes of Might & Magic* gets a very low score for efficiency, having some seven different kinds of resources that are needed to produce things. Further, the effectiveness of a particular unit was extremely hard to determine due to *stacking*. For instance, an individual skeleton might be a rather weak creature, but it's very hard to gauge how tough they are collectively if there's a stack with 117 of them.

Advance Wars is a good example of a clean, elegant design (particularly the first Game Boy Advance release and the later reboot *Days of Ruin* for the DS). There are maybe ten different unit types and each has its own special role—and that's it. The game is very simply about configuring various military units on different kinds of terrain (about five or six). Despite a few small flaws, *Advance Wars* is a good example of an elegant game design.

Take Advantage of Each Decision

If you're designing a turn-based game, you have a wonderful setup for a very efficient, interesting game that has literally no downtime. Each decision, each move, is an opportunity for a really interesting, meaningful, ambiguous decision. If your game has a grid, consider reducing its size. If you are controlling units, consider making fewer with more power. Remember that each element that adds complexity blurs the decisions in the game slightly, and this can add up quickly: before you know it, you've got a game with 10,000 nearly meaningless decisions instead of one with 100 very interesting ones.

Grid Design (or Not)

Squares are the go-to grid setup for most turn-based game developers because they are perceived as being the easiest for people to understand or use. This is unfortunate, because squares bring a lot of problems.

One problem with squares is that diagonal movement takes players farther than orthogonal movement does. Some systems charge players movement points or some such thing for moving diagonally, but this is something of a Band-Aid solution. For this reason, war gamers tend to go with hexes (hexagons). Mechanically speaking, the strength of hexes is huge because each tile is equidistant from all of the tiles surrounding it. It's a nice, uniform system and I would recommend it as the default replacement for squares in turn-based games. The only two downsides to a hex-based design is that it can make it more difficult to incorporate nice-looking artwork into the game, and also many control setups (keyboards and D-pads) are based on a four-directional control scheme. Of course, the latter isn't a problem for mouse-based or touch screen–based games.

Another thing to consider is that even though grids (the splitting up of *space* into discrete sections) generally do make sense in turn-based games (since the element of *time* is divided up into discrete sections), you don't always have to go with a grid. The tactical RPG *D&D* module *Temple of Elemental Evil* (developed by Troika Games) gives you a continuous amount of movement to spend on a continuous space. Using continuous space instead of a grid requires a little finessing, but it can be done well and might be right for what you're doing.

Role-Playing Games

- Examples: *Final Fantasy, Fallout, Ultima, Mass Effect, Wizardry*

In almost every game you literally play a role. In fact, playing a role—or having agency in the world of the game—is kind of what makes a system interactive to begin with. When speaking about the genre of role-playing games, though, I'm referring specifically to games that are heavily thematic and story-laden with leveling up, inventories, parties, and usually, turn-based combat. As you can tell, I think that there's a lot wrong with RPGs, and some of the suggestions I make next to improve them may cause them to no longer be RPGs. But first, a little background is in order.

In many ways, the RPG genre is one of the most important genres in video-game history. For one thing, video games owe their existence in large part to *Dungeons & Dragons* players who used that game as an inspiration to create many of the world's first computer games. Further, the complicated statistical systems and large amounts of bookkeeping in RPGs really exploit the digital platform, since these elements would otherwise have to be done by hand. The computer made RPGs much easier to play than they ever were before!

But RPGs (and *D&D* before them) are also prime culprits in video games becoming fantasy simulators. Almost everybody loves fantasy, and further, fantasy is something we understand. Games are abstract and hard to fully comprehend, but put me in a dungeon with a sword and a shield and let me fight a dragon, and what don't you get? RPGs became the adult equivalents of playing cops and robbers or house—a way for people to explore fantasy worlds. That aside, however, most early RPGs were largely hack 'n' slash dungeon crawlers. Most of them were somewhat hard to understand, requiring a thorough reading of manuals and other documentation, and usually involved lots of hot keys. Many of them were crushingly difficult as well.

RPGs originated during the 1970s and 1980s in the United States, largely because the United States was the world's leader in computer technology then. But by the 1990s Japan had begun its contributions to the genre as well. The JRPGs have their own distinct take on the genre, putting a very strong focus on presentation. They have some really fantastic music—the *Final Fantasy, Dragon Quest, Saga Frontier*, and *Mother* series are all known worldwide for the music. *Chrono Trigger* was famous for its artwork by Akira Toriyama, the artist from the *Dragon Ball* anime/manga series. They also put a very strong focus on a completely linear story, and are highly playable, requiring only a small amount of instruction to play, in part because they are played on consoles like the NES. Today, all RPGs are very story-based, and while the US RPG has largely left the scene, JRPGs (and games inspired by them) are still extremely successful.

Games Hurt Stories, Stories Hurt Games

If you are tying your gameplay to a linear story, you are causing huge damage to your game. If you are allowing the player to change the course of the story through gameplay, you are causing huge damage to your story. There is simply no way around this, so if you *absolutely must* combine the two, you should probably choose the one that is more important to you. There is simply no way to combine the two without damaging one or the other.

If your primary goal is to tell a story, you should consider another medium, such as a short story, a comic, or a screenplay. The idea that pressing a button to see the next page means you're more immersed is pure myth. We become immersed in any activity that flows well, and this can happen in any medium. Contrary to popular opinion, video games are not better at immersion just because you're holding a joystick. An audience will become immersed just as much in a great film, album, or play.

Quicksave the Destroyer

The whole thesis of this book is that games are great when they force players to make difficult, interesting, and ambiguous decisions. If you allow players to save a game-state right before they make decisions, then you're pulling the rug out from under that. A decision has no weight—and worse, no ambiguity—when all possible routes can be tried out in a matter of minutes.

Of course, people are going to say, well, what's the alternative—we can't make people play through our boring game *again*! The implications of these kinds of motivations should be obvious. Moreover, I'm not saying that you can't have any kind of game-saving feature. An inoffensive alternative for saving games is a save-and-exit style mechanism, which could also be called *suspending the game*. This type of mechanism simply lets players save their games when they want to stop playing—loading is only possible from the title screen.

Randomize

As I said at the beginning of this chapter, randomizing your game is the only way to create any replay value in a single-player game. If you randomize the gameplay, you can actually allow players to *lose* games when they make bad decisions. It might seem impossible to randomize an RPG and have it still make sense, but it's definitely possible. Look at games like

Dokapon Kingdom or *100 Worlds Story: Tales on a Watery Wilderness* for very basic examples of how to start, and expand from there.

Does Leveling Up Make Sense?

As a game progresses, it should get harder, not easier, since players are getting better at playing the game as they go along. If the challenges don't increase, it means that after players have overcome one obstacle they don't need to be too concerned about the rest.

The classic idea of leveling up, however, increases your character's stats—numbers that determine your character's attack power, speed, durability, etc. This necessarily means that on some level the game is becoming easier, not harder. Of course, developers are wise to this, and they increase the difficulty even more to compensate for the player's additional power. This becomes a difficult two-axis balancing act, one that becomes essentially *impossible* when you factor in the fact that the player's leveling will be irregular. Not all players will level up at exactly the same rate, or in the same way, and so the idea of balancing such a game is insanely difficult. Resolving this issue generally comes down to making the gameplay slightly too hard, and then just letting players grind as much as they want. Players then have to basically guess at how much they should be boring themselves until there's a challenge ahead that's roughly balanced. This "solution" is totally absurd and any game that functions this way represents a total failure in game design. Players should *never* have to do the job of the designer.

A High Fantasy False Choice Is Still a False Choice

Theme is no justification for false choices. If the player is doing something that is uninteresting, it doesn't matter if thematically, he's single-handedly saving the world from dragons; the game will be uninteresting. You can love the world, the characters, the music, and even think that the game has some cool mechanics, but if you're being asked to perform brainless chores, your brain can't help but see them as brainless chores.

A great example of this is the overmap in JRPGs. The overmap pretends to be a continuous space, but really it's just a discrete space. You could easily represent the overmap from, say, *Final Fantasy VII* in a menu, like this:

* [Grind]
* Midgar
* Sephiroth's Barn
* Chocobo Castle
* Boss Monster Cave
* Menu

Other than these few discrete choices, the rest of the overmap *means nothing*. This makes the task of finding those discrete locations, and walking to them, nothing but a chore. Grinding in particular is an especially annoying chore, since it's nothing but a random amount of time between random encounters. Instead, you should just have a button that takes you right into a fight, as on my menu. (Of course, this does seem a little stupid. But that's only because random encounters are kind of stupid to begin with, and this is simply shining a light on that stupidity and making it easier to see.)

Sports Games

* Examples: *John Madden Football, Ice Hockey* (the NES game), *Mario Tennis, Mutant League Football*

In a way, the genre of sports games is like a brother to the board-game genre of war games, in that most of them are simulations. Digital sports games usually fall into one of two categories: games that are more original or abstract that are loosely based on a sport, and games that strive to *simulate* a sport. Most frequently, games aren't trying to simulate the sport itself, but instead the televised version of it. For instance, probably the most famous sports-game series in the United States is the *John Madden* series of football games (now just called *Madden*). Since their humble beginnings in 1991, these games have done everything they can to look like TV football (the early Sega Genesis games even digitized audio clips of John Madden speaking, although the audio quality was extremely low). The games have come a long way in terms of technology, and it's fair to say that they look a lot like real TV football.

By now some readers will be able to predict my next question: so what? Why is it so important that football video games look like TV video games? Well, if we're judging them as games, it doesn't matter. So are there really enough people who want to recreate the magic of watching a football game on TV for this simulation-style game to become this popular? War games represent a niche in the board-game world, because despite the fact that many historical battles are fascinating and interesting things (unlike watching something on TV), there's a very limited audience for simulation. So why does the Madden series have such popularity? I think there are many reasons, including the fact that TV football is one of the most popular American pastimes. But I also attribute it to game shame—people simply don't respect games as an acceptable way to spend their time, and so the more a game looks like something else (even TV), the more acceptable it is.

At the other end of the spectrum we have games like *Mutant League Football*, *Crash 'n the Boys*, and *Super Dodge Ball*. These games take an existing game and flex it, bend it, and change it into whatever form they want. The end product sometimes only remotely resembles the original game. And of course, you've got some games in between these extremes: *NBA Jam* takes the TV-style basketball game and puts a crazy superpower spin on it (the *NFL Blitz* and *NBA Street* series do a similar thing). And then you have more down-to-earth games like the NES games *Ice Hockey* and *Mario Tennis* for the Nintendo 64, which change their games in subtle ways to make better video games.

Free Yourself from the Sport

If you want to make a simulation, then go for it, of course, but this is not a book on how to make simulators. If you want to make a game, you should commit to that. Committing to making a great game means not caring about whether your system is 100%, 50%, or 0% similar to a particular sport, whether it's being played in real life, on TV, or in a house. Sports were designed for a very specific set of controls: the human body moving in real space. You aren't working with those controls. Your controls are probably buttons, or touch screens, or something else. So it only makes sense that your game's rules would be different—and quite possibly *very* different—than those of the real sport.

Here's the good news: by freeing yourself from the guidelines of the real sport, you're free to make a game that's *even better than the original sport*! Even more good news: because we've never had a very strong philosophy of game design before, it's actually not incredibly hard to make a game that's better, as a game, than most sports. If you look at the history of US football, you'll see that we're really still shooting in the dark about what the sport even is. You can have a principled, strong design, right out of the gate—an opportunity that real football will never have.

Consider Perspective

OK, so you're going to make a soccer game. But from whose perspective? Putting a player in the role of a forward is going to make for an extremely different game than if he or she is playing as a goalie. In most sports games, you control all of the players, but maybe in your game you just control one. Maybe you aren't a player at all, maybe you're just a coach—playing as the coach could still be a very interesting game, particularly one that emulates US football with its plays system (it would be a bit like real-time, phase-based chess or *Robo Rally*). Or maybe you're none of these roles, maybe you're the manager (as in the *Football Manager* series). Considering perspective should be a fundamental thing, and not an afterthought. Nothing should be taken for granted in game design.

Racing Games

- Examples: *Gran Turismo, Super Mario Kart, F-Zero, Excitebike*

So-called racing games are in large part contests. For the most part, all players are making exactly the same decisions in each part of a course, and those who excel do so largely because of memorization and execution. The more these games are purely about racing, the closer they are to contests. Even in pure racing games, however, there are definitely some ambiguous decisions to make in terms of what the other racers are going to do. Since you and other racers cannot occupy the same space at the same time, you'll sometimes have to fake out other racers or try to predict what they'll do in order to get ahead of them (or to stop them from getting ahead of you). Other than that, there's not much room in pure racers for ambiguous decision making, which is why they fall largely into the category of contests.

As with some other genres, we have a primary spectrum in racing games between the literal, simulation-like racing systems like *Gran Turismo* and more abstract systems like *R. C. Pro-Am*. Many games fall somewhere in between, such as *F-Zero* or *Excitebike*. Much of my criticism of racers is similar to what I said earlier in this chapter about sports games, and anyone designing a racing game should review that section.

A Course with Choices?

Many racing titles have shortcuts—secret, hidden, or simply hard-to-get-to routes—which are usually the best way to go. The only reason players would not take them is if they felt they might mess it up, so it's too much of a risk. You might think that choosing between the longer route and the shortcut is an ambiguous decision, but it isn't. The optimal move is taking the shortcut. There may be some ambiguity about whether to try it at the beginning of the race, but the decision is fairly cut-and-dried, and before long, you'll be trying to take the shortcut every time.

Instead, consider having different routes with distinct advantages and disadvantages in your design. Maybe one is longer, but gives the player an item. Maybe one is much shorter, but causes damage to your vehicle. Those kinds of choices make racing a lot more interesting.

Items and Other Features

The *Twisted Metal* series of games are not racers, exactly, but these games offer a lot of examples for using asymmetrical forces and special power-ups to turn racing contests into games. *Carmageddon*, which was known for its graphic violence, was actually quite experimental in a lot of ways. Not only were there special power-up items and stuns that you could perform for bonus points, but there was also a terrifying, invincible, AI-controlled "police truck" (something of a monster truck combined with a construction vehicle) that would hunt the racers down. Adding this kind of a side threat to a game otherwise focused on racing created some very interesting situations that forced players to make tough decisions.

Much Shorter Courses

I recommend that game designers try making racing games with tiny tracks. In general you should start as small as possible with anything you design, and then move up in size only as needed. A small racecourse means that players will be interacting with each other more frequently, even if one player is much better than the other.

You Can't Fix the Skill Deficit Problem...

This harkens back to our discussion about difficulty in Chapter 3. A multiplayer game often must address the problem of skill deficits—if one player is significantly better than another, the thinking goes, it just won't be fun for either player. If the skill deficit is actually so large that one player has no chance of winning and the other has no chance of losing (outside of throwing the match), then in a way it's not a contest, and therefore not a game.

The problem of skill deficits tends to be a big one in racing games, which depend on high levels of both skill and memorization. I'm sure that most of you have had something like this happen: you buy a cool new racing game, and start playing it. You play it and play it, beating various single-player challenges, and so on. So you tell your friend, "I'm really excited about this new racing game I got. You should come over and play it with me!" And your friend does! But only minutes after your first game, you realize that even though this game is great, playing it with your friend is not great for either of you. You don't see each other or interact with each other in any way—you may as well be doing a single-player time trial.

Developers have realized this for a long time, and their solutions boil down to two things. One is to add items to the game. Items didn't start with so-called kart racers—they had been around in games such as

Rare's *R. C. Pro-Am* and Blizzard's *Rock n' Roll Racing*.[1] But kart racers—most notably, the *Mario Kart* series—were the first to award items with extremely high levels of randomness. In the *Mario Kart* games, you're practically guaranteed to get an item every time around with no extra effort. The items are placed right in the middle of the track, and it's much more rare to *not* get an item than to get one (it's difficult to avoid getting them!). The items are given randomly, so this adds a tremendous luck factor to the game—some items are very powerful and take you straight to first place, while other items are quite weak or situational. Of course, this "fixes" the skill deficit, but at the expense of rewarding good play.

The other "solution" is rubberbanding—a feature that slows down players near the front and gives players near the back extra speed. Rubberbanding is often exaggerated near the finish line to create a dramatic last minute neck-and-neck win moment. What developers fail to realize, though, is that that kind of a moment cannot be manufactured.

You can fool people a few times with these techniques, but when players don't feel like they have earned their victories, that's a problem. You're messing with absolutely the most fundamental part of games: building skill. In a game that's very random or with strong rubberbanding, you excel whether you're good or bad—so what reason is there to push yourself? What motivation is there to use imagination? What would inspire someone to use creativity? Games have to be tough, and they have to be internally consistent and fair, because without these conditions human beings just won't care.

Then how do you solve this problem of players having different skill levels? Well, there's an underlying attribute of racing games that makes this somewhat unfixable.

...Unless You Make a Game Out of It

As I mentioned before, the reason that the skill deficit issue is such a problem in racing games is that racers tend to be closer to contests than they are to games. Which is to say, they are (to a large degree) a simple exercise in measuring which player has the better skills, whether that skill is knowing exactly how to best handle the steering mechanism or having the level memorized. In a pure racing game, if your opponent knows these two things better than you, the only way you will win is if he or she messes up. In a game you may be able to throw your opponents a curveball and surprise them, but in a traditional racing contest there's nothing you can do to win if you're losing.

---

[1] A more recent release, New Star Games's *Super Laser Racer*, seems to carry on in an older item tradition.

So how do you make a racing contest into more of a game? Well, I made some suggestions already, but hopefully your goal is clear: to create some kind of system that allows ambiguous decision making. Remember to start from scratch. If you take nothing for granted, chances are you'll discover some amazing new form of racing game (or driving game, at least) that has never been imagined before.

Consider Handicaps

The idea that one player will be handicapped in order to make a game more competitive is extremely unattractive to most gamers, especially digital gamers. This is unfortunate because it is an incredibly useful and good tool for doing exactly that. Instead of taking the Nintendo route and simply making decisions matter less for everyone (which I think is throwing the baby out with the bathwater), you can simply turn on handicaps. It might be useful to realize that handicaps have been used for thousands of years and is actually an inherent rule in competitive Go playing. There is no shame in playing with a handicap, or against a player who is using one.

Fighting Games

+ Examples: *Tekken, Street Fighter, Mortal Kombat, Super Smash Brothers, Virtua Fighter*

A direct ascendant of the world's oldest game—*real* fighting—a fighting game is usually a one-on-one combat game, often themed with two humanoid characters punching and kicking each other, and usually seen from a side view. As with most other genres we've talked about so far, fighting games tend to come in two main subgenres, with a few outliers. One thing that's common to all fighters, though, is that they're highly asymmetrical, with various characters each having its own set of special moves, strengths, and weaknesses.

The first subgenre is the 2D fighter. This genre exploded in the 1990s with *Street Fighter II* and *Mortal Kombat*, as well as a dozen spin-offs, sequels, and other games that tried to emulate these two games. Some might go so far as to say that *Street Fighter* gave us what we now recognize a fighting game to be. 2D fighters are still in production as of this writing, although they have declined massively since the 1990s. They often have health bars (or meters), and later games introduced other bars such as super bars. They're also known for complicated input sequences that players must memorize and execute in order to do special moves. Some 2D fighters (one of the early ones being *Killer Instinct*) became known for a system called combos—moves that would lead into other moves. Often, once you've hit a person with the first move the rest of the moves in the sequence proceed automatically.

In the late 1990s, we got some of the very early entries in the 3D fighter subgenre, with *Tekken*, *Virtua Fighter*, and a few others. Some credit an earlier PlayStation game, *Battle Arena Toshinden*, with being the first truly 3D fighter, but the aforementioned games were the ones to make the subgenre popular. In these games, *side stepping* into the *z*-axis became an element of gameplay. This may sound like a small change—and in the larger scope of things, it actually is—but if you had only played 2D fighters, it was monumental. Now, a spinning kick would still hit you even if you moved to the side, but a straight-on jump-kick would miss. Further, many of these games added hit-you-when-you're-down abilities—which of course, could be dodged by rolling out of the way. These small changes, which made players think about which way they were going to go, added a lot of inherent complexity to 3D fighters.

In response, 2D fighters started becoming more and more complicated themselves. As of this writing, many fighters are released with upwards of thirty or forty characters, four or five special bars that characters need to fill up and spend during gameplay, and *thousands* of special moves and inherent rules that a player must learn to really play. There also have been a couple of outliers, such as *Super Smash Brothers*, *Rag Doll Kung Fu*, and *Power Stone*, which started from scratch and asked fundamental questions about what these games were going to be.

Fighters are special in that they have always maintained their *game* status; indeed, it would have been pretty hard to lose it because of the fact that real fighting always has been and always will be a game. Today, we also see massive communities of professionals playing *Street Fighter* competitively. Along with FPS games and RTS games, fighting games are some of the most-played e-sports—that is, professionally played video games.

Again, Consider Symmetrical

A really fascinating thing about fighting games and the professional fighting-game community is the obsession with asymmetrical forces. One would think that if these were truly great games, that they would still be great games even with just one character (i.e., symmetrical forces). Yet, most people seem to think that the symmetrical matches in *Street Fighter*—also known as mirror matches—are the most boring parts of these games. How could this be? Could it be that when you peel back that extra complexity, the core gameplay isn't all that strong? If a house becomes worthless when you remove the furniture, what does that say about the house?

Why are all fighting games asymmetrical? Who made up this rule that they all have to have 12 or more characters? Why can't they just have one? This should have been tried by now, at least. But, if you are going to make an asymmetrical game, do me this favor: play with it as a symmetrical game for a long time while you're testing it. The game must stand up on its foundation, just like a house. Towards the end of development, when you're confident that the core mechanisms are strong, you can add the asymmetry.

Too Much Complexity

If you decide to use asymmetry, don't go overboard. Start with a small number of characters—something like three or four—and see how deep and flavorful you can make those characters. If you have a great new idea for another character, then try it out. Start at three characters, and move up, slowly and only when you are truly inspired to do so. Do not, under any circumstances, choose a number of characters up front and then try to meet your quota of 8, 12, or 20 characters.

Further, consider the number of moves very carefully. Keep in mind that these are real-time games, with continuous movement onscreen. That means that even if you had only two moves—like, say, a jump kick and a block—the amount of emergent complexity is quite large. People take for granted the immense amount of gameplay meaning and information passed along by a real-time game with continuous space. So, again—start with just two or three moves, and then increase the number *only as needed*. Do you really need a "strong and fierce" punch? Do you really need both an "uppercut in place" and a "jumping uppercut"? All of these add more inherent complexity, which your players must learn before they are able to really play, so you're making the game harder to play and harder to balance. Don't overwhelm your players with noise moves—it makes any good moves your system had to start with lose their identities.

Health Bars

Most fighting games have two colored bars at the top of the screen representing each character's health. When a bar is completely empty, that player has lost. There's nothing wrong with health bars, but the fact is that they aren't tied in to the game happening on the screen all that well. They have zero relationship with the player's onscreen position, and they don't convey much about the match itself. All that matters with a health bar is whether it is depleted or not; no other state has any effect on the game, and for this reason it's somewhat flat.

It may be hard to see that there are other alternatives, but the N64 game *Super Smash Brothers* is a fantastic example of what's possible. In this game, like in other fighters, players maneuver and fight each other on a 2D plane. In this game, though, the designers decided that *positioning* would be everything. The primary role of attacks is to knock enemies back and ultimately to knock them off the stage. The game's levels are designed with platforms and various places to jump around, which emphasizes the positioning element of the game even more. Finally, the health system in *Super Smash Brothers* is totally different from the one usually found in fighters: instead of losing health, players actually gain damage. As their damage increases, the distance that attacks knock them back also increases, thereby tying the health system directly to the core mechanism of positioning.

This was one of the only times I saw a developer deviate from the classic health bar thing, but it is by no means the only option. Think outside the bar—there are a thousand possibilities out there, waiting to be discovered.

Too Much Input Complexity

In the 1990s, in arcade fighting game circles, I remember how cool it was to know all the characters' moves. This came to a head with the release of *Mortal Kombat*, which had finishing moves completely unrelated to gameplay that you could perform with a complex series of button presses. If you knew all the Fatalities, as they were called, you were the coolest.

Well, it's cool when you're 12, anyway. The idea that designers would purposely make a move more difficult to input than it has to be is completely senseless (Figure 20). Unless you want people to not learn how to play your game, don't do this. Designers could get away with it in the 1990s when games were few and far between. Now, there is such incredible access to games that if you do something offensively dumb, such as "do a half circle twice and then all three punches to execute your super move," people will simply move on. It shouldn't be a badge of honor for players to just be able to input commands into your game. It should be a badge of honor that they make good choices during play.

Semi-Lofty Spinning Head Butt (Delft Blue Level)

Figure 20. The semi-lofty spinning head butt—first cousin to the haymaker squat punch in Chapter 2.

FPS Games

- Examples: *Doom, Halo, Quake, Team Fortress 2, Battlefield, GoldenEye 64*

First-person shooters are the games I probably have the most experience with. As early as 1994, I hooked up two PCs (a 486 and a 386) by way of their soundcards' serial ports in order to play *Doom* multiplayer. I played a lot, and these led to some of my earliest forays into some light game design in the form of level designs for *Doom*. Since then, I've always had an FPS game icon (or several) on my desktop ready to go at a moment's notice.

As with the other genres, two major subgenres have developed. One is the high-damage precision-based games with more realistic guns, such as *Counter-Strike* or *Day of Defeat*, and the other is an older, slightly more arcade-like style with high health, fast movement, and imaginary, exotic guns (i.e., *Quake, Halo,* and *Unreal Tournament*). The latter style is often a bit more of a simulation, oftentimes loosely simulating a real conflict. In fact, the US Army created an FPS game called *America's Army* that is very much a literal simulator.

Besides these two subgenres, there are also FPS games that are more creative and game-like. If you read the sections on sports games or racing games, you can probably already tell which of these I think is the better route for a game designer. But I'm getting ahead of myself here.

Ditch Silly Conventions

The FPS genre is a genre of many silly conventions. Among these are health and weapon pickups, armor, and respawn locations. The health and weapon pickups are small boxes that are located in specific places around the map that either give you health or a new weapon. They're stupid to different degrees depending on the game, but they're almost always bad. In old deathmatch-style games, weapons were found in specific locations. Often, these locations would be camped (kept under surveillance) by a player so that another player couldn't get to them. The items respawn after about 30 seconds or so once they're picked up, so what can happen is you'll be in a duel with another player, and the health or weapon just happens to respawn next to your opponent while he or she is near it. The other player picks the health up and wins the match.

Now, it would be one thing if the game was about positioning: moving each other in such a way so as to move your opponent out of the range of the pickup. But this is instead a vestigial design element that's there just because it is. It should be rigorously questioned before going into your game design.

Beware the Sniper

I know everyone loves snipers, but this is a great example of something that players might love but might actually not be in their own self interest—or to put it another way, in the interests of the game. Snipers are *inherently* overpowering, and their nature (long range, high damage) is such that they come with other problems too.

The problem is, *either* of those characteristics would be enough to justify the class. A class with long range and low damage would still be really usable, and a class with short range and high damage would also be (and is) really usable. So snipers are kind of too powerful inherently, and their level of power should be balanced out by a really, really strong weakness (for instance, maybe they can't move for ten seconds after they fire). The weakness that would be required would probably be greater than the level that someone would tolerate.

Consider this also: if you're playing as a sniper, part of what makes you a sniper is that you have an especially long range. That means that most of the time you, as a sniper, are simply fighting other snipers.

Team-Based? Do Something about It

Since the early 2000s, team-based shooters have really taken off. This is largely due in part to the phenomenal success of the *Half-Life* modification, *Counter-Strike*. One good thing that *Counter-Strike* did for team games was to add voice chat, so that teammates could communicate with each other in real time. Before that, players had to stop and type out messages to one another, which is totally impractical in a high-stress situation.

Developers need to do more of this sort of thing. One avenue that hasn't been explored much in terms of this kind of improvement is removing personal scores. If soccer players were allowed to see their personal stats ranked on a big board against those of their teammates in real time, for instance, it would probably screw up the game. Players would start doing things that were good for their personal stats, but not necessarily good for their team. But this is how it works in today's popular shooters such as *Team Fortress 2*. Players have two interests that often conflict—personal score and team score—and the game does not do a great job of highlighting which is the important one.

In the case of team-based shooters with character classes, such as the aforementioned *Team Fortress 2*, I'm actually going to compliment asymmetry because it serves to promote teamwork. When each player on the team has a unique role, it means that players have to work together more. Unfortunately, *Team Fortress 2* has several loner classes (the Sniper, the Soldier, the Spy, and the Scout are examples) who really can hold their own without interacting with any teammates. Some FPS designer should take this to the next level! Maybe there are only three classes: one can shoot, one can heal, and one builds infrastructure. Something like that—use your imagination, but know that each class has to have really strong and distinct weaknesses to make sure that the other classes are needed.

A Third-Person First-Person?

More recently, there has been a swell of "third-person" FPS games. (Of course, this doesn't make sense, but designers do it anyway.) Essentially these are first-person shooters, but for reasons not related to gameplay these games show players their characters from third-person perspective (*Gears of War* is an example)—when players aim or fire, the game uses a sort of over-the-shoulder camera mode and zooms in a bit more.

The issue here is that large sections of the screen are being taken up for no good reason. Why do we need to see the character again? Why is seeing a 3D model of the back of this character (whom I cannot interact with) more important than seeing the enemy who may, at a given moment, be *behind* that 3D model? The only reasons given for this are nonsensical statements about immersion, or feeling tied to the character. I refer anyone who says this to Chapter 1 of this book. Unless hiding information from players is a game mechanism, you should never be placing large solid objects in their field of view. Just make the game first-person.

Avoid Single-Player

First-person combat does not make for very interesting single-player gameplay. The reason for this is that the mechanics of aiming at something and shooting is actually, on its own, not terribly interesting. It's flat and largely an execution contest. The ambiguity and stimulation of these games comes from trying to read the actions of an opponent.

Unfortunately, AI-controlled enemies tend to be extremely predictable and therefore uninteresting. For this reason, I can't advocate making a single-player FPS *unless* it's something radically different than anything we've seen before—something that doesn't use aiming and shooting as the core gameplay mechanism. But then . . . is it really an FPS anymore? For now, I would say that unless you have some revolutionary idea for a randomized, score-based FPS that somehow makes shooting really interesting (perhaps a fast-paced roguelike FPS could work?), go for multiplayer.

Platformers

- Examples: *Super Mario Brothers, Banjo-Kazooie, Sonic the Hedgehog, Spelunky*

Platformers are real-time video games that involve navigating an avatar through space, usually jumping from platform to platform. A central theme of platformers is that if you fall off a platform or miss a platform on a jump you fall to your death, and either lose a life or the game. Part of me didn't want to address platforming games in this book, but the popularity of *Super Mario Brothers* and hundreds of other titles it inspired forced my hand.

My reason for not wanting to include them is that there's kind of only one game involved—the game is one in which you jump from platform to platform to get to the end of the level. Simple as that! Many games have added new features or some new spin to the core mechanism, but in a way, you can say that all of the games are really different expressions of the same core game. Then again, perhaps that could be said of most video-game genres—or even all genres in any medium! But I think that this characteristic is a little bit more pronounced in platformers.

As with the other genres, the platformer genre has splintered into two subgenres with the advent of 3D graphics. With Nintendo's *Mario 64*, millions were introduced to the idea of a platforming game that was fully 3D. This dramatically changed the nature of the gameplay in platformers, which I'll get into more next.

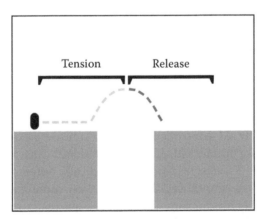

Figure 21. An illustration showing where tension and release are found in a platforming jump. Tension is felt up to the point of no return, which in *Super Mario Brothers* is roughly at the apex of the jump. The more control you give players in midair, however, the more the area of tension shrinks.

Platformers generally work by having a very clear pattern of tension and release: the jump. The jump is their core mechanism. When you jump, the tension builds as you are rising to the apex of the jump. You don't know for sure whether you launched yourself with the correct trajectory, but the answer starts to become increasingly clear during the flight. While this is happening, tension forms in anticipation of what may or may not be about to happen. By the peak of the jump you can usually tell whether you're going to make it onto that next platform. Once you land, the tension is released. This pattern of tension and release is the primary engine by which these games work (Figure 21).

Protect the Jump

Just about every platformer gives players at least some leeway to change their trajectories while in midair (there are some exceptions, such as the earlier *Castlevania* games). However, you have to be extremely careful with this feature, because if you give the player too much control you destroy the core mechanism of the game. If I can just "fix" my jump completely while in mid-air, how can there be any tension? Remember that tension comes from anticipation, and if I have complete control at all times, there is nothing to anticipate; everything is immediate.

In *Super Mario Brothers*, players had a certain amount of ability to change their directions in midair during jumps. In *Super Mario Brothers 3* (the next true *Super Mario Brothers* game to come out in the United States, as the US version of *Super Mario Brothers 2* was actually a completely different game called *Doki-Doki Panic*, reskinned with a *Mario* theme), this ability increased. With the sequel that followed—*Super Mario World* for the Super Nintendo—the amount increased again, to the point that you could almost completely undo an inaccurate jump.

These two sequels also added many items to the game, the most important of which were the Super Leaf from *Super Mario Brothers 3* and the Cape Feather from *Super Mario World*. These items both allowed players to dramatically increase the level of control they have over their jumps, along with allowing them to fly (which we'll get back to in a moment). Later versions added other features that further interrupted the jump, causing tremendous damage to it. The biggest offender was *Super Mario Sunshine*, which gave Mario what was essentially a jet pack that let him hover at any time, until he lined up perfectly with where he wanted to land. In this game, the jump was completely ruined as an engine of tension (except for the levels that didn't include the jet packs, which many people said were their favorite levels).

The latest entries in the series as of the time of this writing are the *Super Mario Galaxy* games. Thankfully they've gotten rid of the jet pack, but now Mario has a double jump in the form of the spin move. This move can be used to correct most bad jumps, and thus destroys most of the tension of the game.

I really can't stress enough how important the jump is in a platformer, and it pains me to see so many game developers—indie developers in particular—misunderstanding this. For instance, Team Meat's lead designer Edmund McMillen said this about his game *Super Meat Boy*:

> It feels to me better than *Mario*, which was in my mind the perfect way for a platformer to feel. It feels like *Mario*, but in a lot of ways, a lot of aspects of it feel better. It feels faster. It feels like I have more control, especially in the air. I feel like I have complete control over the character, and that is…number one with a platformer…

To me, this is yet another example of the "more is more" philosophy creeping in. McMillen is saying that the controls in *Super Meat Boy* "feel better" than Mario because the player has more control. He even uses the phrase "complete control," which implies that he doesn't understand how the jump mechanism works at all. Since game design is all about carefully choosing limitations (rules), game designers should never be bragging about giving players "complete control."

Flight? Really?

I remember my excitement as a young child when I heard about Mario's new ability to fly in *Super Mario Brothers 3*. How fantastic! Mario can now soar through the air like a bird! Wow! Well, now that I'm an adult I can see how counterproductive this addition actually was. Adding flight to a platformer makes as much sense as putting a racing game on rails, or adding auto-aim to a first-person shooter. It's allowing the player to completely ignore the core mechanism of the game.

People think it's exciting to fly up and over an entire level or large parts of one. Sure, for a second it's exciting in the same way that entering in an invincibility code is exciting, but as anyone who has used such a code knows, it gets old quickly. It is not robust: it's flat and the initial thrill of being able to basically skip the game wears off quickly. I mean, why stop at skipping the level? You can skip the whole game by not even playing in the first place!

Randomize

I know I mentioned this at the top of this chapter, but the lack of randomization is really a serious problem for these games. As of this writing, there's only been one marginally well-known platformer that has randomized its levels: *Spelunky* for Windows (apparently an Xbox Live Arcade version is on its way as well). There need to be more.

The issue is that after the first time you play through a platformer level that's not randomized, you begin to memorize it and the level begins to get solved. After playing a *Super Mario Brothers* level just a few times, a player will pretty much have it memorized. As I mentioned at the top of this chapter, memorization means that with each play, your skill is being tested less and less.

3D Platformers Are a Bad Idea

I've stated that there are problems with almost *all* third-person 3D games, but the issues are particularly pronounced with platformers. The reason is that in a platformer your spatial positioning is absolutely crucial information: pixel-perfect precision in jumps can mean the difference between life and death. Yet in a 3D platformer, players have to translate so many different angles without the use of depth perception in order to make jumps that more often than not, they miss jumps simply due to missing information.

Figure 22 shows a 2D platforming situation: the amount of distance that the player has to cover is quite clear. All of the important information is there—there is no guesswork involved in determining precisely how far away that next platform is. The player can say with absolute certainty that the gap is roughly three character-widths wide.

Figure 23 shows a similar situation in a 3D platforming game. Now, can you tell me how far away that other platform is? The answer is, you really can't. The best you can do is estimate—make a guess. It could be three character-widths, but it could also be two or four. It's also possible that it could be a mile away and absolutely huge.

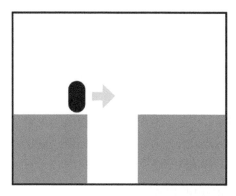

Figure 22. Perspective in a 2D platformer.

Figure 23. Perspective in a 3D platformer.

There are also other problems, such as camera angles, with 3D platformers. Figure 23 shows the viewpoint once the player has corrected the camera angle (which again, the player should never have to do if the game designer has done his or her job). But the player has several different angles to consider even after the camera angle has been corrected.

- The angle between the direction your *platform* is facing and that of the next platform (this matters if you need to run forward before you jump).
- The angle between the direction your *avatar* is facing and that of the next platform (this can matter with some special types of jumps that launch you forward without running).
- The angle between the direction your *camera* is facing and the next platform (this is often useful so that players can hold a cardinal direction, such as up, on the joystick to have *some* reliable leverage point).

I think that this is just way too much for the player to have to calculate. More importantly, none of it has anything to *do* with platforming. This may change once we have 3D screens that can give us a really good sense of depth perception. But for now, if you're developing a game for a 2D output device, keep the gameplay 2D if you can.

Other Genres

In this section, I'll quickly go over a few other genres, using our lens to make a few important points about them.

Point-and-Click Adventures

- Examples: *Myst, Leisure Suit Larry, Maniac Mansion*

While point-and-click adventures have many merits, game design is not one of them. For the most part, these are not games at all, but puzzles. Worse, they are very often *bad* puzzles. They are notorious for hunt-the-pixel problems and other totally arbitrary problems that must be solved. All of the merits of such games—without any of the annoying frustrations—could be enjoyed by watching a person play the game on YouTube.

Interactive Fiction

 ♦   Examples: *Zork, Adventure, Facade*

See the "Games and Story" section in Chapter 2 for better insight into why putting *interactive* and *fiction* together is simply a bad idea. Marrying them makes designers choose between having a bad story, having bad gameplay, or having both suffer a little bit. It's not completely impossible to make something good in this genre, but it is a bad idea to try to do so when there are other media that are better for stories and better for gameplay.

Shooters and Shmups

 ♦   Examples: *Galaga, Gradius, Contra, Metal Slug, Ikaruga*

While there's nothing inherently wrong with shooters and shoot-'em-ups (shmups), their play tends to break down into memorization puzzles. As far as I can see, there are no random games in this genre. However, some of them have some very interesting gameplay mechanisms that could really be expanded upon, such as the *Treasure* games *Bangai-O* and *Ikaruga*. These are also sometimes referred to as bullet-hell games, because of their habit of putting hundreds of bullets onscreen. I find that this term tends to understate the real genius of these games, however.

Abstract Puzzle Games

 ♦   Examples: *Tetris, Bejeweled, Puzzle Quest, Dr. Mario*

Abstract puzzle games are absolutely not puzzles at all, but they still get called puzzles. I attribute this to our habit of calling anything that isn't an action game a puzzle, or perhaps to the fact that the "pieces fit together." Regardless, these games have a lot of potential and need to be explored more: being abstract means that the gameplay is always the focus. I highly recommend people experiment more in this genre, because there's another *Tetris* waiting to be found right around the corner.

Roguelikes

 ♦   Examples: *Rogue, Nethack, Diablo, Titan Quest*

In 1980 a PC game called *Rogue* was created, which spawned a niche genre that is still alive today. In fact, it may be more popular now than it ever was before. That genre is called roguelikes.

People argue about what makes a roguelike a roguelike, but generally, the genre follows a few conventions: random map generation, turn-based play, a single controllable unit (a character, as opposed to a party), score-based play, and permadeath (permanent death). Many roguelikes have ASCII symbols instead of artwork, are set inside dungeons, are highly influenced by *D&D*, and have very high levels of inherent complexity.

Roguelikes are a great example of what single-player games could be if they were randomized, but they are also a great example of the excesses of the digital medium, often having thousands of arbitrary pieces of information that players must know in order to excel. I highly recommend that any game designer become at least marginally familiar with some of the best ones, such as *Mystery Dungeon: Shiren the Wanderer* (NDS), *Dungeon Crawl: Stone Soup* (Windows/OSX/Linux), or perhaps even my own *100 Rogues* (iOS/OSX).

Video "Games"

To round out the chapter, here are some categories of video games that aren't really games at all.

MMOs

- Examples: *World of Warcraft, Ultima Online, Guild Wars, Eve Online*

MMOs (massively multiplayer online video games) are not games. MMOs usually *include* games (such as a raid in *World of Warcraft*), but they also contain purely social activities, narrative bits, and many other things. To call an MMO a game is as silly as calling an amusement park a game.

With that said, there is no reason why the games inside MMOs cannot be great, but by placing a great game inside of an MMO you're dramatically limiting your audience. Also, you're usually going to run into the problem of the metagame taking over, which tends to happen in any system that has a heavy amount of metagame (such as RPG elements).

Skinner Boxes

- Examples: *Farmville, Diablo, World of Warcraft, Pokémon*

Some have said that use of the term *Skinner box* is inappropriate or somehow misleading with regard to something like *Farmville*. I disagree—I think that *Farmville* works precisely the way that Skinner's operant-conditioning chamber works. These applications exploit human evolutionary needs and create loops of compulsive behavior, such as the need to collect and gather (clearly exploited and even made into the subtitle of *Pokémon: Gotta Catch 'Em All*), or the need to show status (exploited in *World of Warcraft* when you show off your fancy new gear to others, or vice versa).

In *Farmville* and other so-called social games, the creators must have been aware of what they were doing. Anytime you receive a power-up, the game asks you to click to get it. Many games just give you something if it's a no brainer to take it, but in *Farmville* you have to actually click to do everything. If that's not operant conditioning, then it's just horrible UI design.

I would never say that an inanimate object was evil, but I would warn people about these kinds of applications in the same way I would warn them about getting involved with gambling or addictive drugs. These kinds of games are very addictive but not particularly fulfilling, because they are only exploiting you, not challenging you.

Toys, Sandboxes, and Simulators

♦   Examples: *MS Flight Simulator, Sim City, Minecraft, Garry's Mod*

Toys aren't games. For some reason this is a controversial statement in the digital world, even though any Walmart employee understands that there's one section for toys and another for games (and even another for puzzles!).

Minecraft, for instance, is just a basic interactive system. People tend to mistake this application for a game because they often add their own goals when they play. They have turned it into a game, the same way you can turn *Flight Simulator* or *Garry's Mod* into a game by adding your own goals. In fact, you can turn anything into a game by adding goals and rules: this addition is the process of game design. So give yourself a little credit—if you made up some rules for how you play *Minecraft*, you took part in the art of game design.

5

Through the Lens: Board Games

I t will be very clear to anyone reading this chapter that I do not see board games and video games as being on the same level in terms of game design. That is to say, board games are largely doing a very solid job of fulfilling their potential, and video games are doing an awful job of fulfilling theirs. This evaluation doesn't come from a personal bias: if anything, I would be biased in favor of video games, since like many of my generation I was raised with them and have great affection for many of the classic video games.

This assessment isn't based on some fundamental property of the digital medium either. Video games are capable of everything board games are capable of; actually, they're capable of a lot more in certain ways. There are still things that board games can do that video games cannot, though. Video games can't yet match the social aspect of board games—their ability to get several people sitting around a table—or even the physical feel of components. At the same time the greatest strength that video games have over board games—their lack of physical limitations—is also their greatest weakness. The very limitations of board games, coupled with the world's newfound interest in games, has promoted a renaissance in board-game design.

Whether you're designing board games or video games, I hope that the analysis of various genres of board games in this chapter is useful.

Note that not every genre of board game appears here, and not everyone will agree with the classifications I have set up. In general, I've included sections on categories of board games that I feel will help us to sharpen the focus of our lens; a few lesser-known genres also are addressed to make people aware of them.

Although I entered my mid-20s being extremely serious about games and game design, I had never even heard of the world of board games then, a fact that I sorely regret. I hope that this book will excite the same passion I now have for board games in readers who are currently only familiar with video games.

The Problem with Board Games

As of the time of this writing, board games do not have the rock-star cultural status that video games do, at least in the United States. Board games are often looked at as dry, difficult to learn, and boring. That attitude comes in large part from the 20th-century success of companies like Milton Bradley and Hasbro in publishing very bad, but very popular, board games. These companies had a pattern of taking traditional games, stripping them of any interesting qualities (if they had any to begin with), and repackaging them with new names.

The most famous example of this is *Monopoly*. *Monopoly* was originally created in 1904 by a woman named Elizabeth Magie Phillips. Much like some modern indie art-games, it wasn't necessarily created to be fun, but to make a point. *The Landlord's Game*, as it was called at that time, was trying to make a political point about land ownership—that the capitalist system drives money *upward* and the rich necessarily get richer from such a system.

I need to take a moment to talk about the modern game of *Monopoly*, which we all know, but maybe haven't analyzed. *Monopoly* is one of the worst rated games of all time on BoardGameGeek.com, and for good reason. This game honestly doesn't qualify as a game for adults at all, who, unlike children, should be able to recognize that they have no agency over dice rolls. *Monopoly* is really just an extremely long version of *Candy Land* with player elimination. You really have no choices to make in *Monopoly*: if you land on a space, you should buy it. The rest of the game is entirely up to the roll of the dice. Worse, the game tends to go on for upwards of four hours, frequently with one or two players being eliminated in the first hour. Most people I know have never finished a game of *Monopoly*; the game simply goes on and on until someone gives up because it's just too boring to play.

The original version of *Monopoly*, however, wasn't like that. When Parker Brothers bought the game in 1935 it changed one of the main rules, which was that when a player landed on a space, there was a round of bidding (an auction) to determine who would get the property and at what price. This rule was essentially the *core mechanism* of *Monopoly*, and it was ripped out of all versions of the game long before any of us even learned what the game was.

Monopoly is only one example, but it represents part of the larger picture that many people have about board games—that they take too long, that they stalemate, that they're just about rolling dice and nothing else. Recently, however, games like *Settlers of Catan*, *Carcassonne*, and *Ticket to Ride* have begun to bridge the gap between games totally based on luck like *Monopoly*, and more serious Eurogames like *Through the Desert* and *Puerto Rico*. But the PR campaign for board games still has a long way to go.

The Downside of Interesting

We (especially if we're seasoned video-game players) can usually jump right in and figure everything out in a video game, and we tend to find tutorials an annoyance. Yes, I can figure out that pressing A swings my sword and B makes me jump, thank you very much. Board games aren't really that way: most that you sit down to play will be a completely new kind of experience.

Of course, we all say that we want games to be innovative and interesting and new. But when a game comes out that actually *is* all of those things, it means we have some learning to do, and this can be difficult for many people. Reading rules is a slow and sometimes painful process, oftentimes made more painful than it has to be due to poorly written rule books. As I mentioned earlier in the book, games are nonlinear, and so it can be very difficult to read through a rule book and have any idea what's going on. Often you'll need to read the rule book, then attempt to (sort of) play, then go back and review the rule book a second time before you're ready to actually play. Lots of people have difficulty mustering the patience, concentration, and energy required to give a game a shot if doing all this is required up front. And, unfortunately, the more new and interesting a game is, the more this is required.

Eurogames and Ameritrash

I don't use the terms *Eurogame* and *Ameritrash* as my categories, because these terms don't describe game mechanisms, but rather regional styles of board-game design. I've included descriptions of what each is, however.

Eurogames are games created by Europeans, most frequently by Germans. Germany has become something of a mecca of board-game design, and it hosts the world's largest board game convention, The Spiel (held in Essen). These games usually have somewhat dry or minimal themes, often about farming, medieval courts, or trading in the Mediterranean Sea. They are notable for two things: first, thematically, they are almost never violent or about war. Secondly, they embrace game mechanisms and the art of game design in a really unprecedented way. In fact, Eurogames are also sometimes called designer board games because the game designer's name appears on the box.

Ameritrash was originally a pejorative term, which has since been adopted by lovers of the style of game and is now in common use. These games, usually created in the United States, tend to have much higher quality production levels than Eurogames. They often have really intricate components, and bright, striking graphic designs. Ameritrash games also tend to use more exciting themes, such as those based on popular science-fiction TV shows, *Dungeons & Dragons*, or the Lovecraft mythos. They are typically extremely dice-heavy and often derive much of their gameplay directly from *Dungeons & Dragons*.

Area-Control Games

- Examples: Go, *El Grande, Chaos in the Old World, Small World, Samurai*

An area-control game is one in which you try to capture more territory than your opponent (they are sometimes referred to as area influence games). Sometimes the mechanism is expressed through a realistic military conflict, but more often the representation is abstract. There's almost always a grid of some type, although the grid not need be square or hexagonal; games such as *Small World* feature spaces that are unequal in size. In some games, such as the popular tile-laying game, *Carcassonne*, the area to be captured is laid down as the game is played. In these instances, the composition of the area that can be captured is not even known to players at the beginning of the game.

Breaking Stalemates

An inherent problem with many of the area-control game designs is that the simple "I capture one of your spaces, you capture one of mine" pattern makes the game feel static, and in some cases can even stalemate a match. Many games use an element of randomness to alleviate this. For example, the *Warhammer*-themed *Chaos In the Old World* uses dice to simulate combat when you're trying to take a tile that's currently occupied by your opponent's forces. In this game, rolling a six (on a six-sided die) is called an *explosion*, and it allows you another free attack—a random development that can swing a battle in one player's favor dramatically.

Obviously, I don't think that's an elegant solution to the stalemate problem. It would be preferable to come up with a way that players can fake each other out: maybe one player could suggest that he or she is going to take tile A but instead takes tile B. Of course, this kind of element couldn't be a completely free action or the game would essentially come down to rock-paper-scissors—but if the costs associated with switching and moving forces are high enough you could have a much more elegant solution on your hands. Randomness is the easy way out.

Long-Term Planning

Due to the aforementioned randomness, it's often completely impossible for players to plan beyond their immediate turns in such games: players simply have to make the best move given the current situation. Long-term planning relies on too many random factors that could go in completely unpredictable directions.

One possible solution to this is to use a *Robo Rally*–inspired system of asynchronous, perpetual motion. In *Robo Rally*, pieces move at predictable rates each turn; their motion can be modified by the players, but only in a limited way. This allows for a larger element of planning, since players are guaranteed a limited number of future game-states. Another solution would be to simply reduce the range of randomness— for instance, instead of ten different types of tiles or cards that can be drawn maybe there are only three.

Bidding Games

* Examples: *Money, Modern Art, Amun-Re, Power Grid,* the original *Monopoly*

Bidding is one of the oldest forms of game playing, as it is an extension of the familiar economic activity of price setting—a very natural and dynamic way of finding out what something is worth. Decisions in a bidding game are often quite difficult to make and involve considerations of how far to press your luck and the degree to which you know your opponent.

Bidding games start with some object of value on the table that more than one player wants. The first player bids a certain amount of resources (usually themed as money) for the item, and the next player has the option to either pass or make a bid higher than the original one. Sometimes bidding rounds allow a player who has passed to come back later and bid, but more often players who pass are out of that bid. In some games each player must pay whatever is bid, but in most only the winning bidder (determined by the highest stated price when all other players have passed) pays.

The engine of this system is based on trying to gauge the value of in-game items: if you can judge the "true" value of an item, you'll know how much money you should spend for it. But the system is deeper than that due to an inherent bluffing mechanism—if you know another player wants something, you can bid a bit higher than you think the item is worth in the hopes that the other player will pay more than he or she wants to for the item. This requires not only knowing what the item is worth, but also having a good grasp of what the other player thinks that item is worth. And of course, the other player can always call your bluff and leave you to pay more than you want to for an item you may not have wanted in the first place.

Controls

One common downside to bidding games is that sometimes it can feel as though the amount people end up paying for things is somewhat arbitrary, or even random. Of course, winning bids are not random at all. Often, though, two players who both want an item badly will cause its price to increase to an unnatural level, leaving a hole in the currency of the game that produces imbalance (for at least one player, but possibly for two or more depending on the game).

It's probably a good idea to put some kind of controls on the bidding. Perhaps there could be only three or four levels of bidding, and players either can't go higher, or when someone does go higher a special trigger is activated that changes the rules in a dynamic way. It's worth mentioning that excessive bidding tends only to be a problem for newer players, as more experienced players will generally be more careful with their money.

Dimension

For bidding games to be interesting, bids must involve a sufficient variety of conflicting attributes. For instance, if you and I are bidding on one victory point, we obviously both want that victory point equally. There is no dimension to this question: while it's still ambiguous as to whether you will bid higher than my current bid, the interplay is flat, shallow, and uninteresting.

But say we have the same situation, except in addition to the victory point that's available for auction there are two other items, one of which you want (perhaps it's the other half of an item you have that will give you five victory points when completed). The more dimension there is to an item that's being bid on, the more information players have to predict their opponents' behavior.

War Games

- ◆ Examples: *Squad Leader, Ambush!, Diplomacy, Axis & Allies, Risk, Memoir '44*

In a way, I didn't want to include war games in a separate genre because *war* is really just a theme, not a mechanism (if anything, the mechanisms of war games could probably be included in my previous section on area-control games). However, I can't simply ignore the tremendous popularity of this genre, and so for the same reason I included platformers in Chapter 4, here it is!

War games are usually (but not always) heavy, long, and complex simulations. They often simulate real-life historical conflicts, with in-game mechanisms that attempt to simulate the way actual military troops, vehicles, and weaponry of the given time period actually worked. For this reason, they are some of the (if not *the*) most difficult kinds of games to get into. In fact, if you don't know someone who's already into war games, the chances of your getting into them is very slim. With that said, war games are highly strategic and tactical games with a tremendous number of interesting decisions to make. They are challenging, yet flexible, and despite their dry presentations they really do allow for creative play. War games usually take place on a board with a hexagonal grid, have various campaigns and scenarios to play out, and incorporate dice rolling into their combat situations. *Advanced Squad Leader* (a more recent version of *Squad Leader*) is probably the world's most popular war game of its kind.

That's a description of the war gamer's war game. There's also a lighter breed of war game, the most popular of which is probably *Risk*. Everybody knows *Risk*, and love it or hate it (most serious board gamers are in the latter camp due to the extreme levels of luck involved in the game), it's likely the world's most well-known war game. You also have some slightly more elegant, more board game-like (less simulation-like) takes on the genre with the *Commands & Colors* system and similar systems. *Commands & Colors* is well-known for its use in the popular war games *Memoir '44* and *Commands & Colors: Ancients*.

Finally, you have some oddballs that are still war games, but don't fit into either of the categories above. One such game is *Diplomacy*, a game with no randomness and rather simple gameplay mechanisms that is played out over a course of many hours. *Diplomacy* is a strategic/tactical game—it's played on a grid, but it's less about the mechanical taking of area and more about forming alliances with other players and breaking them at the worst possible times. Many have called *Diplomacy* a friendship-killing game for this reason.

Why isn't chess a war game? Well, it is a game that at least loosely simulates war, so it wouldn't be incorrect to call it a war game. However, we'll be addressing chess and other abstract games in the "Abstract Games" section of this chapter.

Forget Simulation

If you want to learn how to make a better interactive simulation, it should probably be clear by now that that's not what this book is about. If you want to make a *game*, then you should do everything in your power to make sure you're doing the best job that you possibly can.

The first thing you have to do is put simulation as a distant second priority to having great gameplay. The sad fact about reality is that reality isn't always balanced. Real-life conflicts aren't always interesting to play out, and real-life weapons and vehicles can be a pain (especially when you make players deal with stuff like refueling and jammed weapons). At the beginning of this book I stated that games can occur naturally. While this is true, I didn't say that *great* games occur naturally. Great games almost always have to be created by a human mind that understands the fundamentals of what makes a great game.

As I've said before, it seems extremely unlikely to me that you'd be able to simulate something in a somewhat accurate way and not hurt your gameplay. Further, even if you aren't hurting your gameplay with a commitment to realism, you are certainly ignoring any nonrealistic possibilities that might be interesting and open your game up. For this reason I generally recommend that you start with an abstract design; once you have great mechanisms in play, then you can try to fit a specific theme over it if you like.

Think Outside the Genre

Many war games play out in a very similar way: you move your units, I move mine, your units attack my units, mine attack yours, you capture this tile, I capture that one. But there's such a huge range of possibilities as to how you can express a war; you don't need to follow the patterns that have been in use since the beginning of the 20th century.

First, look at area-control games for inspiration. They are a fantastic resource for interesting expressions of movement, methods for resolving combat, and more. Abstract games also have plenty to teach the war-game designer.

Consider using something other than a hex or a square grid—maybe a triangular grid or an irregular grid would work better for your game. Perhaps players could lay down the tiles as their troops are moving across the map, like *Carcassonne*. There are thousands of possibilities out there. Study the fundamentals of the war-game genre, but stay on the lookout for game mechanisms that are used outside the genre as well.

Role-Playing Games

+ Examples: *Dungeons & Dragons, Shadowrun, Paranoia, Call of Cthulu*

Video-game players should note that here, role-playing games constitute a completely different genre than the RPGs they're familiar with. Also called *pen and paper RPGS* or *tabletop RPGs*, these are games that are played with several people sitting around a table using dice and a rule book, sometimes using a grid and figures for tactical combat, and often using the imagination of a *game master*.

The game master is arguably the most interesting part of this breed of interactive system, although also frequently a trouble area for it. The game master (called the Dungeon Master when playing *Dungeons & Dragons*) is sort of like a real-time game designer. First, he either chooses or creates the campaign that the players will go through. Then, during the game, he takes the part of any monsters or other opposing forces. He also is the storyteller of the game, setting the scene and describing what areas look and sound like.

Are these systems games? Well, it really depends. The original version of *Dungeons & Dragons* was heavily inspired by war games that came before it, and early versions were more competitive (creator Gary Gygax dreamed of *D&D* being played competitively, and there are even first-edition modules that were designed for tournament play). But modern *D&D* is a huge mix of so many things—fantasy simulation, game, interactive storytelling, and pure social activity—that it makes it difficult to call it a game.

With that said, there are some systems that are more (and less) competitive. John Harper's *Agon* is a quick, competitive system that I probably would classify as a game; same goes for Atlas Games's *Rune*. In contrast, there are also systems like *Sorcerer* and *Dogs in the Vineyard* that are almost entirely about storytelling and have very few mechanisms.

Look Beyond *D&D*

If you're new to pen and paper RPGs, it might seem like *Dungeons & Dragons* is a good place to start. I would advise against this in most cases. The reason is that *D&D*, now in its fourth edition, has a lot of baggage. The game has always been somewhat unfocused, and you're probably not getting the best bang for your buck in terms of how much you're going to learn versus the materials you'll have to buy and read.

In the past decade or so, there has been a very steady rise in independently created pen and paper RPGs. I recommend doing some research, buying or downloading some PDFs, and playing with some of these systems (check out BoardGameGeek's sister site, RPGGeek.com).

Emergent Stories

As I said earlier, the interactive story is always at war with itself. I won't say don't make an interactive story—if that's what you want to do, go right ahead. But as I also said earlier, it's unwise.

Keep in mind that there's a lot you can do with pure games that is social and has a lot to do with human interaction. The traitor game *The Resistance* is fantastic at creating emergent stories, most of the mechanisms of which are simply people's ideas about each other, rather than pieces on a board or cards in a hand. All games (and all activities, actually) create stories, but social games such as *The Resistance* take place largely in the verbal realm. I recommend doing something like that, with the feel of a story-based game yet not tied down to a linear narrative.

Cooperative Games

- Examples: *Pandemic, Forbidden Isle, Arkham Horror*, Reiner Knizia's *Lord of the Rings*

Closely related to solitaire games, a cooperative game is played by multiple players against the system itself. These games use randomness—usually in the form of card draws, dice rolls, or both—to create adversity and simulate an opposing intelligence. While cooperative games tend to be beloved by more socially oriented gaming groups, the genre has some inherent flaws that thus far no one has been able to completely solve.

Mistakes Are Good

The relationship is actually *too* close. If you want to, any cooperative game can be played in single-player mode simply by controlling all of the players on their turns. In and of itself this is obviously not a problem; in fact, it's kind of neat that it's an option. The problem is that if one person has a much higher skill level in a cooperative game, he or she will tell everyone else what to do. This happens all the time in cooperative games; one person basically takes a leadership role and figures out what the game plan will be for the next four or five turns. A lot of times, what that player is telling everyone to do is at least a decent call—often a *better* call than what the other players would have come up with.

Some think that this is really not such a terrible thing. The truth is, if you're into these games merely for the social aspect it isn't so bad. The damage done to your game experience, however, is tremendous. A huge part of playing games is creative exploration, and a huge part of creative exploration is being able to make mistakes. If you aren't allowed to make your own mistakes, then you really aren't playing at all. So how do we resolve this issue?

I actually had a daylong conversation about this issue with three other game designers at a recent game designer's conference. We figured out that there were essentially only two ways to fix the problem.

- *Make the game competitive-cooperative.* Competitive-cooperative games are games that either one player will win, or all players will lose. Therefore, some amount of cooperation is required to avoid total failure on all sides.
- *Add a traitor mechanism to the game.* The issue with purely cooperative games is that there can be no hidden information. For instance, *Pandemic*'s rules tell players not show other players their hands. However, players do need to know what cards the other players have, and if they can't show each other, they continually have to ask instead. It's just a silly rule. Adding a traitor mechanism sets up the possibility that one player might *not* be an ally, though, which creates a real, in-game motivation to not reveal cards.

What you may have noticed is that with either of these solutions, the game is no longer purely cooperative. I feel comfortable saying that the potential lack of creative exploration in cooperative games is a difficult problem that no one has solved yet, and it's possible that it may be unsolvable. It would be great if someone could come up with a completely new way to play that resulted in the cooperative game surviving when played with players of drastically different skill levels.

Role-Selection and Worker-Placement Games

- Examples: *Puerto Rico, Agricola, Caylus, Citadels, Dominant Species*

Role selection and worker placement are not the same mechanism, but they're close so I figured I'd hit them both at the same time. Typically,

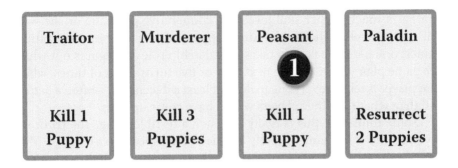

Figure 24. An example of how role selection works. Depending on the game, players select role cards by either placing a token on a card or simply taking the card.

role-selection games allow you to temporarily (often, just for the current turn) choose a role. Usually others can't choose the role that you took for the duration of that turn, so there's often a bit of a race to pick a certain role. The role you choose allows you to perform a certain action or gain a special power (Figure 24).

Worker-placement games are somewhat similar, but sometimes allow players to place more than one worker in a given "role" (for lack of a better word). For instance, suppose a game has a mining box and each token a player places in that box means that the player will get one more gold piece at the end of the turn. Sometimes worker-placement games allow multiple players to place their workers in a given box, but in other games a box may be considered off limits to all but the player with workers in it.

These two mechanisms are extremely flexible and can be (and are) used as core mechanisms for thousands of different types of games. *Citadels*, *Puerto Rico*, and *Age of Steam* are all very different games, each with a different theme and different mechanisms, yet they all use a role-selection mechanism to express their basic gameplay.

Consider Dynamic Roles and Actions

There's not much that I have to complain about with these "genres," but I do have one possible suggestion for those looking to create such a game: consider making the roles themselves change dynamically throughout the game. I've seen examples of games that sort of do this (like card games in which certain cards are designated as available each round, but the specific cards that are available change from round to round—*El Grande* is one example), but I've never seen something that exactly fits the bill of role selection. The idea that you can increase the effectiveness of one role for everyone, not just yourself, is something that would possibly open up a game to all kinds of new dynamic, interesting decisions.

Card Games

- ♦ Examples: Poker, Rummy, Crazy Eights, Contract Bridge, Cribbage, *Tichu*

Probably *every* genre I've listed in this section utilizes cards. Here, I'm defining *card games* as those that use a 52-card deck (sometimes called a *poker deck*). Within this classification there are several other subgenres: partnership games, betting games, solitaire games, children's games, and many more.

There are literally thousands of different games that can be played with the same deck of 52 cards—it's fascinating how much flexibility this system has. For this reason, a deck of cards is a great tool for teaching game design (I use card decks when I teach game design to children). There are also some games, such as *Tichu* and *Haggis*, that use a minor variant on a 52-card deck. The fighting card game *Yomi* also uses a 52-card deck, but with a lot of extra information added.

Be Suspicious of Convenient Numbers

If you're developing a card game, and it just so happens that it works with exactly 52 cards, in four suits, with three types of face cards, etc., be wary. In short, what are the odds of this actually being the optimal setup for your game? It's like a unit in *StarCraft* with exactly 1,000 horsepower; you can't help but feel that the amount of power might just have been a ballpark guess, as opposed to the exact balanced level of health for that unit.

First, experiment with a half deck. Experiment with two or three decks. Experiment with taking out face cards. Experiment with adding in 17 completely new, unique cards that you made up. Playing cards are convenient, but don't let them hold you back.

Abstract Games

- ♦ Examples: *GIPF* game series, *Arimaa*, Go, Chess, Checkers, *Hive*

In some ways abstract games are the best games you can play, hands-down. Their almost total lack of theme allows the mechanisms to be extremely rich, deep, and elegant. The absence of theme also means that there's a pretty firm cap on how inherently complex an abstract game can be, because it's theme that helps us understand more complex systems.

Chess is known all over the world and has been played professionally for hundreds of years. Massive numbers of thick books have been written about chess strategy, and famous games have been memorized and scrutinized. The game Go is like chess in this respect, but even more so—it has been played for literally thousands of years (over 4,000 at least) and it's been solved to a much smaller degree than chess.

Many people don't realize it, but a lot of new abstracts come out all the time. In fact, there are free websites that frequently publish new, online versions of abstract games for you to check out (one of my favorites is BoardSpace.net). I also recommend looking into homemade print-and-play abstracts.

It's always harder to create something simple that's also interesting. For this reason, abstract games tend to be the most difficult games to create. Further, they tend to be hard to market, as a lot of people—sadly—won't give abstract games a chance due to the lack of theme. Hopefully, as people become wiser about the true nature of games being *inherently* abstract, this will not be the case in the future.

Avoiding Solvability

In 2002, Omar Syed and Aamir Syed developed *Arimaa*, a two-player abstract game that was designed to fix what they saw as the problem with chess. The game's creation was inspired partially by the famous chess game between top player Gary Kasparov and Deep Blue, the chess computer, wherein the computer was victorious. The concern was that chess had been solved, at least partially. And it's true: computers *have* partially solved chess.

Whether or not this development affected the world of chess, or even whether a potential full solution for the game would affect that world, isn't really the point. The point is that abstracts do have a tendency to become solvable. I used chess as an example because although it is not one of the more solvable abstracts, solution looms on the horizon for even that deep, fantastic game.

Many games have already been solved, such as the popular abstracts checkers and *Connect Four*. Once a game has been solved, this obviously causes huge problems for it. Of course, actually *using* a solution is sometimes quite a process, depending on the game. But the fact that someone *could* use the solution makes competitive or online play somewhat impossible. You could still probably do a *Words with Friends*–style online play, in which you only play against friends on Facebook, but a solved game lives with a harsh cap over its head.

How can you avoid making your game solvable? Well, obviously including random elements means that your game is no longer *hard solvable* (meaning it can still technically be solved with *perfect play*, or using a mixed strategy that makes optimal moves based on probability, regardless of what the other player does). Extreme levels of play testing are the best way to get the low-hanging fruit solutions out of the way, of course. Unfortunately, there's no consistent way to prevent your game from being solved; instead, it's a matter of making sure that that solution is as far out on the horizon as possible. This usually involves adding some layers of complexity to your game.

Theme

If you're designing an abstract game, there may come a point where the game gets a little bit too complex to still warrant being completely abstract. At this point, you may want to add a limited theme—just enough to make the actions, verbs, and tokens more clear to players. Putting too much of a theme in a clearly abstract game can come across as silly to players, and I've heard designers such as Reiner Knizia get criticized for just slapping on a theme.

It's good to realize that a theme is there for *functional* reasons (clarity), not aesthetic ones. An abstract game can be every bit as beautiful as a heavily themed one; in fact, I'd say that Go is more beautiful by far than any Ameritrash game I've ever seen. Add theme only as needed by the game.

Room for Creative Play

Another possible downside to abstract games is that their low level of complexity can sometimes mean it's more difficult for players to play in an expressive way. As a game's level of complexity scales up, one of the upsides is that there becomes more room for play; more room for players to make more moves that don't necessarily seem optimal directly, but which aren't bad either. Lateral, strange moves that can confuse other players or simply express a style of play can be difficult in some very simple abstracts. Of course, this is only something to keep in mind when designing such games, and not at all a problem inherent to abstract games. After all, all games become abstract in the minds of players once they are in the act of playing.

Other Genres

Here are a few popular game genres that don't fit into the categories already described.

Party Games

- ♦ Examples: *Apples to Apples, Trivial Pursuit, Saboteur, Pictionary, Boggle*

Party games are often silly games. They are often not games you would want to play ten times in a row, or for months at a time. They are often distractions, a mere vehicle for fun social interaction.

They don't have to be, though! There is no reason that a party game can't be an excellent game in its own right. All a party game really needs is for the mechanisms to be extremely simple—they need to stay out of the way of the player's social interactions. It's best if the game even uses players' social interactions, as in the case with *Apples to Apples* or *Saboteur.*

Collectible Card Games

- ♦ Examples: *Magic: The Gathering, Pokémon Trading Card Game, Yu-Gi-Oh*

Collectible card games, or CCGs, are fundamentally flawed and can never be great games. Games in this genre have different kinds of gameplay, but it doesn't really matter how great the gameplay is because the very fact that the game's cards are "collectible" means two horrible, balance-destroying factors are in play.

The first problem is that the game is going to have way too much content. In fact, often these games are in a state of perpetually adding new content, which is clearly something that's impossible to balance. The second problem is that not all players will come to the table with the same amount of power. Some players start with different cards than others (based on cards they've bought), which in itself can be an inherent disadvantage. This is compounded by the fact that most CCGs have *rares,* or cards that are better than other cards but hard to find. Also, in an unsavory attempt to exploit their fanbases, most CCG manufacturers sell their cards in packs with a random mix: you buy a pack of 20 or so cards, but you don't know up-front what cards are inside.

In short, CCGs care very little about being serious games and are basically built to take advantage of people. New ones may continue to get made, but nobody will be playing a CCG once its moment in the culture has passed.

Deck-Building Games

- ♦ Examples: *Dominion, Puzzle Strike, Thunderstone, Eminent Domain, Quarriors*

In a deck-building game, you start out with a number of cards in a deck. (Instead of cards, *Puzzle Strike* uses a bag of cardboard chips and *Quar-*

riors uses dice, but I'll use *cards* and *deck* throughout this section to describe the mechanisms.) Each turn you draw a small number of cards from the deck, and then you can usually buy cards that get added to your own deck, and can later be randomly drawn. Some cards increase your buying power, and others give you special abilities. There are also often victory point cards of some kind that give you victory points at the end of the game, but have no other use, thereby diluting your deck. Because of this, deck-building games involve a perpetual balancing act that can make them feel interesting.

Of course, the question of whether or not they actually *are* all that interesting is another one entirely. The issue with deck-building games is that they're extremely luck-based, and often have a pattern of assaulting players with tons of extra expansion content. In fact, I think their mechanisms are actually *not* terribly interesting on their own, so they require this extra content in order to *seem* interesting.

Dominion, which was released in 2008, is the most popular and successful deck-building game. At the time of this writing, there were six expansion packs for the game. Because of the large amount of continually released expansion content, some have taken to calling *Dominion* an LCG, or limited card game. The implied difference between a CCG and an LCG is that there is a limited amount of content available for a limited card game. Of course, this idea is asinine, because the amount of content for *Magic: The Gathering* or any other CCG is limited too. Further, the expectation of continually releasing expansion after expansion seems roughly the same with CCGs or LCGs; the only difference seems to be how the games are marketed (which is notably better for LCGs, since at least you know what you're getting).

Many have hit *Dominion* with the criticism that it's multiplayer solitaire—several players sitting around playing the same solitaire game and then comparing who has the highest amount of points at the end of the game. Then again, *Puzzle Strike* is a much more interactive and direct-attack type of game, and it has been hit many times with the complaint that the game is simply too random for this kind of competition. While you can do some things to mitigate the chances of a bad draw, many good players simply die suddenly because of a bad draw at a bad time.

As I said about games that tend to be asymmetrical: it's not that these mechanisms cannot make for great games, but the fact that they seem to *need* tons and tons of content as well as a flow of *new* content says something about these games on a fundamental level.

Traitor Games

- Examples: *Battlestar Galactica, The Resistance, Werewolf, Mafia, Saboteur*

Adding a potential traitor to a group of cooperative players prevents one player from dominating the rest, since that player may be a traitor who leads all of you into disaster! In traitor games, players all get a randomly dealt role card that tells you whether you're a traitor or not (or the thematic equivalent), and that determines your role for the rest of the game. The only issue with traitor games is that the traitor element tends to overwhelm the other mechanisms in the game. Then again, there are a relatively small number of such games, and many of them are light party games, so that could be the reason for that tendency. I'd like to see a game that embraces this by basing all its mechanisms on identifying traitors, but one that uses a more dynamic system that keeps the traitor role in flux throughout the game.

6

Predictions

W e can make better things if we focus on fundamentals. This goes not just for games, of course. I hope that people can find inspiration in this book not only for the field of game design, but in other creative fields, too.

In most other artistic fields the fundamentals are already established, so if you're a visual artist, musician, or screenwriter you probably already understand a thing or two about the nature of your medium and what makes it tick. Even if you don't, there are tons of books on these media that can help flesh things out for you. You can study your field in school and come out with a deep, working understanding of your medium. For game designers, however, it's not like this.

I wanted to write this book because we are at a prime point in history for this situation to change. And it's not that it *may* change; it definitely *will* change. In the next two decades, we will finally begin to develop some solid guidelines about the nature of games and game design. Many have already begun to call the 21st century the Ludic Age, implying that games will be the defining cultural (and perhaps economic) driving force of the century. Games—or at least interactive entertainment of some kind—are achieving saturation levels never seen before. And yet it has been less than 100 years since *anyone* has been paid to design games full time.

Things are going to change, but how? What will be the driving forces leading us into the next generation for games, digital or otherwise? Much of this book has been negative, but I have very few good things to say about the current digital game industry. I am extremely optimistic about the future, however, and think that many of us are. We live in an amazing time.

The Resistance

Although I think it will be overcome, it must be stated that there is a tremendous amount of resistance towards progress in the world of game design theory. The discussion has essentially remained the same for the last ten years, and the things that have changed it were not sound observations, but software making millions of dollars. The discussion has changed because of *Farmville* and *Minecraft* and *Portal*—not because someone made a solid, bold point about what games really are. Because of this, you'll hear a lot of people say that talking about games is pointless or a waste of time. In a way, they're right, but only because of this resistance to change.

The resistance has a few weapons, all of which are logical fallacies. Any design theory that leads us to the conclusion that *Metal Gear Sold*, *Ocarine of Time*, or *Final Fantasy VII* are anything other than the perfect classics we've always considered them to be is completely off the table. Essentially, we are a generation of people who have a strong nostalgic attachment to the largely broken and dysfunctional game designs we were brought up on. We've absorbed so much abuse that we expect it, and we've committed so loudly and proudly to our video-game patriotism that we can't question it.

The Media

Just as a democracy has the fourth estate to help keep government in check, an art form needs critical analysis and discussion to stay relevant. Sadly, games journalism is currently extremely dysfunctional.
When it comes to professional games journalism, you essentially have four types of writing.

* *Advertising.* A recap of the industry talking points on the back of the game box.
* *Propaganda.* A restating of the current industry-fueled narrative about what's exciting. "What's exciting" may be the new motion-jiggle accoutrement, 3D viewing cables, or whatever other non-game-related garbage that the industry wants you to buy.

+ *Incompetence.* Many of the people doing the writing for sites like IGN, GameSpot, and 1UP seem not to be experts on the subject of games. In their writing they display huge gaps in their knowledge and understanding not only of game genres but also simply of how games work.
+ *Superficiality.* Insightful game reviews simply don't exist in the mainstream. Reviews are shallow, often contradictory, and very rarely have the courage to actually come out and say something. Most game "reviews" are limited to a description of what a game is and a numbered score at the end.

On the other hand, however, we have bloggers: independent people who write just because they're passionate about games. Just as indie gamers will be the ones to lead us into producing new games, bloggers will be the ones to lead us into a new journalism of games. We just need to collectively decide to drop the resistance and allow progress to happen.

Rise of the Indies

A lot has already been written and said about the rise of the independent developer. Most of us who pay even limited attention already can probably name between five and ten independent game developers who came out of nowhere, released a game on Steam or the iPhone, and are now well-known with full-time careers making games that they designed. The indies, it seems, have begun to seriously compete with mainstream, AAA (big budget) game developers and publishers.

Developments

The rise of indies obviously is partly due to platforms like Steam, iOS, Android, PSN, and Xbox Live Indie Games that make it possible for developers to get their games out so that new audiences can easily see and play them. These platforms were necessary for indies to thrive, but it would be a mistake to think that their inception is the larger catalyst behind what's happening in the industry.

Another important (yet less recognized) reason for the indies' increasing ability to compete with the mainstream is that people have gotten very comfortable spending money online in the last decade due to services such as iTunes, Amazon, and PayPal. For instance, even if Steam had been around in the 1990s, the service probably wouldn't have taken off—people simply weren't comfortable with the whole idea of putting their credit card information out on the Internet (and probably for good reason). Since the mid-2000s, however, the percentage of money spent online versus in traditional brick-and-mortar stores has been edging more and more towards online purchases. Buying online is easy, and services such as Apple's App Store make it extremely easy to buy games. The rise of online purchasing was a huge step in terms of helping indies.

The largest factor may be this, though: the culture has changed. In the 1990s, being a game *developer* wasn't cool yet. Being a game *player* (a gamer) already had a certain mixed glamour then: it was a cross between being a badge of coolness and a source of comments driven by game shame, such as "I'm such a geek" (geek, by the way, was more of a pejorative term in the 1990s).

I should share some of my own perspective here, because back in the 1990s I bounced between wanting to be a game developer and a rock star. I was a guitarist, drummer, and songwriter, and I probably don't have to tell you that it was *very* cool to be in a band, especially in the 1990s. It was also very cool to tell people you wanted to be a rock star; people were interested and impressed by this information. When I told people I wanted to be a game designer, though, they just didn't get it. They usually thought that I was saying I wanted to be a computer programmer and kind of tune out. At that time, most people didn't have a well-formed idea of what a game developer was. For something to have a coolness factor, people need to see and understand examples of it.

But that has begun to change. Largely due to social media such as Facebook, Twitter, YouTube and the web itself, game players are in touch with game developers. We retweet their tweets, we link to their articles, we subscribe to their blogs. Not just indies, but the majors, too. We know the names of famous game developers: Hideo Kojima, John Carmack, Todd Howard, Peter Molyneux, Gabe Newell, Shigeru Miyamoto. In the 1990s only the most hardcore gamers knew any developers by name; now millions can identify the names and faces of all of the people I just listed. When you tell someone you want to be a game developer now, those are the faces, personalities, and stories they see. The recent documentary *Indie Game: The Movie* is one of the first major examples of independent developers being introduced to a whole new crowd of people. We're sure to see more of this in the future.

In short, video games are the rock and roll of this generation.

Control

One of the most important aspects of the rise of indie developers has been that slowly but surely, the death grip that major publishers like Electronic Arts, Activision, and Take-Two have had on the industry for much of the 1990s and 2000s is loosening.

In 2001, Sir-Tech released the final entry in its famous *Wizardry* series, but since it belonged to what was considered a niche genre (a turn-based RPG) it was very difficult to get the game on store shelves. Apparently the big publishers felt that such games were out of style, and as a result Sir-Tech (now defunct) had to make a deal directly with the (also defunct) Electronics Boutique game store. Unfortunately, the deal wasn't enough to keep them afloat and the company went under, but one developer from the team later said that if Steam had existed at that time, Sir-Tech might still be alive today.

Because of platforms like Steam, as well as independent sources of funding such as Kickstarter and The Indie Fund, it's easier for developers of all kinds to get their games made and published. No longer are the big publishers the only ones making the calls—the playing field has been leveled.

With people who love games more in control and middlemen less in control, we have even more reason to be optimistic about the future.

Downside

The upsides of the rise of indies are clearly dominant, but is there a downside? There's only one I can see: as much as I hate the institutionalized quicktime-event-action-vampire-cutscene-spam games of the major developers, the fact is that they do have a system. They have teams of 80 or more people sometimes, and everyone knows very precisely what their roles are. There isn't a lot of bickering or debate about what a game will be, and when it's finally decided, few people are surprised by the result. It's either the *third-person action game, reskinned*; *the first-person shooter, reskinned*; *the RPG, reskinned*; or some such thing—there aren't too many options. Indies, of course, have no expectation of following this pattern. If anything, they have an unspoken responsibility to do anything *but* follow this pattern. If that's the case, though...what *do* they do? What will indies do with their new level of power and control?

Currently, I don't think most indies have a very strong idea of what they should be doing with it; most of them seem to be doing one of three things.

- *Recreating retro games.* Many developers simply create games that copy both the strengths *and* the weaknesses of games from an earlier time. The indie team Iron Tower Studios has been trying to create a "last-gen American style RPG" for nearly a decade as of the time of this writing. Spiderweb Software has been doing the same ever since that style of RPG was current. A message to these developers: we didn't do it right back then. Those older games have just as many flaws as the new ones, and if you copy them, you're just copying old flaws.

- *Emulating the majors.* Many indies are trying to create the next *World of Warcraft*, the next *Call of Duty*, or the next *Half-Life 2*. Apparently, these indies thought that the issue was that the major publishers weren't producing enough titles.
- *Making puzzle-platformers.* I don't have the exact data, but given the number of indies currently creating platformers, and specifically *puzzle*-platformers, the majority of indie games being produced must fall in these categories. Somehow—perhaps because of the popularity of *Super Mario Brothers*—the idea that "in video games, you jump around the screen" has been allowed to thrive. Puzzle-platformers seem to be the go-to indie game at the moment.

The indies will be taking over the world of games, and they need guidance—badly. Obviously, this book is an attempt to do my part, but where else might this guidance be found?

Merging Worlds

Changes in technology and the closer association of those who design and play games—no matter the industry—have significant potential to blur the lines between the worlds of video games, board games, and sports. In fact, this melding of worlds is already taking place.

Board Games

There is surprisingly little overlap between the world of video gamers and that of board gamers. As I've mentioned before in this book, I spent the first 25 years of my extremely hard-core video-gaming life playing games and scouring the Internet for obscure games that I may have missed. I always thought that *board games* meant just *Monopoly*, *Sorry*, *Apples to Apples*, and chess—I had no inkling of the fact that there was a whole world of designer board games for me to discover.

Sometime in the late 2000s, though, that started to change. *Settlers of Catan* bumper stickers started appearing. Forum threads talked about *Carcassonne* and *Dominion*. Soon there were webcomics—a cultural bastion of the video-game world—mentioning *Pandemic* and *Battlestar Galactica*.

These days a serious surge of board-game designers and publishers are porting their famous board games to the App Store on platforms such as iOS. Hits such as *Tigris & Euphrates*, *Samurai*, *Puerto Rico*, and

Through the Desert all have digital versions available. Websites such as TouchArcade.com (a website for video gamers to find out about new video games coming out on iOS) review some of these games, exposing a whole new audience to them. The web's board-gaming mecca, Board-GameGeek.com, reached 400,000 registered users in 2011, with growth rapidly escalating. *Carcassonne* was even released on Xbox Live Arcade.

Suddenly all these video gamers are finding out about this other world. They are experimenting with some of these games and I can't help but expect that their reactions will be similar to mine: this is what I have been missing. The opportunity to make interesting, difficult, and ambiguous decisions that I can't take back. The opportunity to explore a game that won't get completed, a game that I can play for years. Isn't this what a game was supposed to be all along—something that I could really *explore*?

Of course, board gamers were well aware of the existence of video games—very few people in the developed world *aren't* aware of the incredible phenomenon of video games. But I think board gamers probably have something to learn from video games, too. The phenomenon of these two worlds merging will be helpful for everyone (although dramatically more so for video gamers).

New platforms will continue to emerge that will facilitate the merging process. One of the most interesting is the Microsoft Surface, which is essentially a very large touch-screen tablet. If these ever become affordable, they will revolutionize the way we play video games *and* board games, and will be a massive step forward in merging the two worlds.

Sports

What about sports? Well, I think the sports and video-game worlds will merge as well. On one side, we already have the increasingly popular pro video games, such as *Street Fighter*, *StarCraft*, and *League of Legends*. These video games function in a very similar way to sports both culturally and mechanically. We also see the same phenomenon in board games: obviously, classic abstracts like Go and chess are played professionally worldwide, but newer games such as *Magic: The Gathering* also have tournament play that's very popular. Many other board-game designers are working on developing tournament play for their games, such as David Sirlin with his *Fantasy Strike* games. It's quite possible that as these board-game and video-game leagues become more and more serious, they may eventually become as popular as physical sports. At that point, it will be hard to distinguish what constitutes a sport and what constitutes a game.

When and if that happens, I think that of the two, sports will have the most to learn. Pro video games and board games are already doing their best to emulate and learn from sports, but the world of sports is still living in the predesigner past, mostly because there is really no way for someone to make a living as a sport designer. However, if there was an opportunity for some smart, experienced, knowledgeable game designers to influence the people in charge of the rules of football, that sport could be in for some very positive changes. Most modern popular sports are the way they are not because of a rigorous process of design, but because of an awkward evolution that involved a somewhat random mixture of good and bad ideas. It's possible that as digital pro sports integrate with traditional pro sports, we'll see more of a focus on game design in sports, something that we currently very rarely see.

Renaissance

All of the factors, all of the signs, all of the trends and advancements are pointing in the same direction: we are heading towards a renaissance in games. We will soon be entering into an era of enlightenment about what games are, why they have value, and how to make better ones.

Why Change Is Inevitable

My prediction of a renaissance is of course not to say that we'll ever have a world in which all games that get made are good. Bad games will always be made, but we'll be much better equipped to distinguish the bad from the good. Games that are celebrated will really be worthy of celebration. They will be games that have lasting value and resonance for people, in contrast to the present day, where many (if not most) of the lauded and celebrated titles are story-based Advent calendars that are discarded once they've been experienced.

Some say that the way things are in 2012 are the way that things will always be. Allow me to list some reasons that support my belief that things will change.

- *It's unsustainable.* The current way that games are produced and marketed is unsustainable and will inevitably lead to either a shift in direction (which is already underway) or a market crash. You cannot perpetually increase the budgets for games every year, especially when the amount players are willing to pay for games is decreasing. You can only trick people into buying the same game, reskinned, so many times (and I mean so many, but it is still a finite number of times). Even if people don't realize that, despite the hype, the games they're playing are boring, people will simply find themselves less and less attracted to the idea of playing games at all. I know many people who love games, but because they only know about digital games, they've essentially quit playing. Many more will follow in the future.

- *Merger with board games.* Board-game designers are years ahead of the curve right now. When video-game designers are exposed to the philosophies of board-game designers (either by playing their games or through direct communication), those who design video games will be unable to ignore their insights. The non-philosophy of video-game design will be seen for the absurd nonsense it is when this new way of looking at *what games are* is shined on it.

- *Increased discussion.* More people are engaged in discussions about video games than ever before. Forums are alive with talk about game design. GDC talks are available for viewing on the Internet. New game design conventions are springing up. More and more books are being written on the topic of game design. In short, we're working on the problem. The only thing holding us back in this regard is the strong anti-progress culture that still exists in most discussion circles. Once we realize that any serious answers may force us to kill some of our darlings—to be a little bit destructive—we will get past this.

- *There is simply no other route forward.* If we don't develop a dramatically better understanding of what games are and how they work, then what is the route forward? The latest thinking seems to be that we need new hardware gimmicks: Nintendo's Wii, Microsoft's Kinect, touch screens, and several other bizarre (and extremely *limited*) technologies have gotten a lot of attention. Some even claim that 3D viewing screens are the answer, although it should be very obvious to anyone who plays games that these things aren't what makes a game great. Great game mechanisms are great game mechanisms, whether played with a Wiimote or an Atari joystick; whether on a monochromatic Game Boy screen or a 56-inch plasma 3D television.

Are there any other serious proposals about what the future of games should be? We can all agree that it's a good thing that more indies are able to make games now, but simply allowing indies to experiment isn't a clear way forward. Some have suggested that "social" games are the future. Obviously, most games are *social*, and the classic examples of *social game* are simply exploitative compulsion engines, more akin to slot machines than anything else. So I ask you: if developing a deeper understanding about what games even *are* isn't the way forward, then what *is*?

Music during the Renaissance

What does a renaissance look like, exactly? This isn't some term I am using carelessly. What we will experience will be a lot like the impact of the Renaissance on music in the 15th century. Until then, Western music had largely been the province of churches, used primarily as a vehicle for various religious ceremonies and practices. Almost all development that took place in music in the Middle Ages resulted directly from Church commissions.

However, in the 15th century—a time of great cultural and technological change—things began to change. The rise of the new bourgeois class, combined with the development of the printing press, meant that for the first time music became somewhat self-sufficient. Music was being produced and listened to *for its own sake*, not just as a means to an end. By the time the Renaissance was over (roughly 1600), a system of functional tonality that all music is based on had been established. (The only exception is the somewhat limited instances of music in which a conscious decision is made to *reject* functional tonality; examples include noise music, post-tonal modern composers, and other avant-garde works).

In short, during the Renaissance we pretty much figured out the basic functionality of music. And this is what's going to happen with games. Right now, video games are still in the phase where games can't really exist for their own sake—most often, they have to justify their existence with some sort of fantasy simulation. Just as this was not the case for music made after the Renaissance, this won't be the case for games made after the Ludic Renaissance.

Purpose

Because it wanted us to continue buying its hardware every few years, the video-game industry has told us the same lie for years: newer is better. *Newer* often took the form of *bigger, more,* or *higher levels of technology,* but by now it should be clear that those things do not make *better* games. Ironically, however, the mantra *newer is better* may actually become somewhat true in the future.

If and when we reach a point of enlightenment about games, the games we make are going to improve dramatically. Our games will become much more coherent, interesting, and lasting than they ever have been before. We'll no longer be shooting in the dark.

There will always be bad and mediocre games, of course. But I hope that this book can be a useful and significant stepping-stone towards a better future. Anyone reading this book will have access, in their lifetimes, to the greatest games that civilization has ever seen. For people who love games, there has never been a better time to be alive.

Index